ROUTLEDGE LIBRARY E[...]
ENERGY ECONOM[...]

Volume 9

THE ENERGY CRISIS

THE ENERGY CRISIS

Ten years after

Edited by
DAVID HAWDON

LONDON AND NEW YORK

First published in 1984 by Croom Helm Ltd

This edition first published in 2018
by Routledge
2 Park Square, Milton Park, Abingdon, Oxon OX14 4RN

and by Routledge
711 Third Avenue, New York, NY 10017

Routledge is an imprint of the Taylor & Francis Group, an informa business

British Library Cataloguing in Publication Data
A catalogue record for this book is available from the British Library

ISBN: 978-1-138-10476-1 (Set)
ISBN: 978-1-315-14526-6 (Set) (ebk)
ISBN: 978-1-138-30481-9 (Volume 9) (hbk)
ISBN: 978-1-138-30485-7 (Volume 9) (pbk)
ISBN: 978-0-203-72974-8 (Volume 9) (ebk)

Publisher's Note
The publisher has gone to great lengths to ensure the quality of this reprint but points out that some imperfections in the original copies may be apparent.

Disclaimer
The publisher has made every effort to trace copyright holders and would welcome correspondence from those they have been unable to trace.

THE ENERGY CRISIS ten years after

Edited by
DAVID HAWDON

CROOM HELM
London & Canberra
ST. MARTIN'S PRESS
New York

©1984 David Hawdon
Croom Helm Ltd, Provident House, Burrell Row,
Beckenham, Kent BR3 1AT
Croom Helm Australia Pty Ltd, 28 Kembla St.,
Fyshwick, ACT 2609, Australia

British Library Cataloguing in Publication Data

Hawdon, David
    The energy crisis: ten years after.
    1. Energy industries   2. Energy consumption
    I. Title
    333.79'12   HD9502.A2
    ISBN 0-7099-1288-9

All rights reserved. For information, write:
St. Martin's Press, Inc., 175 Fifth Avenue, New York, NY 10010
First published in the United States of America in 1984

Library of Congress Cataloging in Publication Data
Main entry under title:

The Energy crisis ten years after.

    Includes index.
    1. Energy policy – Congresses. 2. Energy industries –
Congresses. 3. Petroleum industry and trade – Congresses.
I. Hawdon, David.
HD9502.A2E54364   1984   333.79   83-40189
ISBN 0-312-25123-8

Printed and bound in Great Britain

CONTENTS

ACKNOWLEDGEMENTS

The success of The Energy Crisis Ten Years After
Conference in April 1983 at which the papers which
make up this volume were given, was due in large
measure to Mrs Jan Benn of the University's Bureau
of Industrial Liaison and to Mrs Heather Platt of
the Energy Economics Centre. The conference
sessions were efficiently organised by Sir Richard
Meyjes, Chairman of Council, Mr A J Wiggins of the
Department of Energy, Professor George Ray of the
National Institute of Economic and Social Research
and Professor David Simpson of Strathclyde
University. Sally Silverman provided a vast input
of secretarial and word processing skill without
which this book could not have been prepared.

THE CONTRIBUTORS

Rt. Hon. Nigel Lawson M.P., Secretary of State for
Energy of the United Kingdom

Professor Colin Robinson, Head of Department of
Economics and the Energy Economics Centre,
University of Surrey

Professor Edith Penrose, Professor of Political
Economy, Institute Europeen d'Administration des
Affaires, Fontainebleau

Sir Peter Baxendell, Chairman, The Shell Transport
and Trading Company

Professor Morris A. Adelman, Professor of Economics,
Massachussetts Institute of Technology

Professor Wolf Häfele, Director,
Kernforschungsanlage Julich, Federal Republic of
Germany

Dr. Herman T. Franssen, Head, Economic Analysis
Division, International Energy Agency, Paris

Ray Dafter, Energy Editor, The Financial Times

Dr. Walid Kadduri, Executive Editor, Middle East
Economic Survey, Cyprus

D.r Robert Mabro, Fellow of St. Antony's College,
Oxford

Michael Parker, Director of Central Planning,
National Coal Board

INTRODUCTION AND SUMMARY

David Hawdon

The international energy economics conference held
at the University of Surrey in April 1983 was
important for two reasons. It was one of the first
opportunities for academic, business and government
economists to assess a full decade of unprecedented
change in the world energy situation since the oil
price increases of 1973. But secondly, and quite
adventitiously, it occurred immediately after the
March 1983 conference of OPEC in London when, with
the tacit agreement of non member countries like
Britain, the official price of oil was lowered
substantially and an output sharing agreement made
between member countries.
     This involvement of governments with energy
was the theme of the first paper, presented by Nigel
Lawson, the UK Secretary of State for Energy, which
was noted by the press as a significant statement of
UK energy policy. He argues that instead of
allowing their countries to adjust to higher world
oil prices after 1973, too many governments had
hindered the process by increased borrowing.
General inflationary pressure was one cause of the
continuing rise in oil prices in the period but this
pressure had now been reduced. One legitimate
objective of government policies was to minimise the
fluctuations in prices which cause disruption to
economic activity whilst another was to reduce
monopoly power in all energy markets by encouraging
the development of a variety of indigenous
resources. UK policy sought to promote stability in
energy markets.
     The economic history of the last decade is
largely the history of the oil market. However,
Colin Robinson placed developments in the oil market
in the context of overall energy trends in the
postwar period. Prior to 1973 the uninterrupted

growth experienced in the western world is seen as a
direct consequence of falling real energy prices as
oil and gas supplies expanded and economies of scale
were realised in transport and refining. In the
same period market forces steadily eroded the
dominant position of coal until by 1973 oil and gas
together accounted for two thirds of all energy
consumed. The sudden changes in the energy market
in the early seventies and particularly from 1973
onwards are explained in terms of standard resource
theory which in its simplest formulation asserts
that producers will expand output so long as
expected rates of price appreciation (dP/dt) are
less than their discount rates (r). But when
discount rates fell with ownership changes in the
early seventies and at the same time price
expectations rose due to fears of resource scarcity
the cutback in oil supplies and the associated price
increases which occurred can be seen as the rational
market response to dP/dt r, irrespective of OPEC or
anyone else. Why then has it taken so long for the
post 1973/78 market to adjust to higher prices in
the post period? Colin Robinson attributes this to
a lack of response of consumer prices to producer
prices, and therefore to a lack of incentive to
reduce energy demands; to increased government
taxation of energy profits which reduced the
incentive to invest in new supplies; and finally to
the inherent time lags in markets requiring
adjustments in capital stocks with a public opinion
increasingly hostile towards replacement energy
sources like nuclear power. This tardiness of
response is seen as the major reason why the world
energy market is likely to remain dependent on oil
and therefore subject to occasional swings in oil
prices for the foreseeable future.
        The role of OPEC is developed further by Edith
Penrose. She argues that although OPEC played a
central role in bringing order in oil markets in
March 1983 through its success in reconciling
conflicts of interest between member countries and
in securing the co-operation of non member producers
like Britain and Mexico; its role in oil markets up
to then has often been exaggerated. This is due to
the essential divergence of interests between
members resulting from unequal distributions of
reserves and population levels. Its existence has,
however, provided a medium through which the
oligopolistic power of such producers as Saudi
Arabia and the Gulf States could be more effectively
exercised. But its future influence is jeopardised

by the entry of new oil producers, the weakness of
the oil companies and the still great gap between
oil prices and costs which invites cartel breaking
behaviour. On the other hand consumer governments
have an interest in protecting their conservation
investments and perhaps OPEC can act as the core of
a producer/consumer grouping with similar
interests. After a careful analysis of the weakness
of OPEC's current position, the conclusion is drawn
that it may be possible for OPEC to preserve a
degree of stability, although prices may fall
somewhat below $29 per barrel.

One of the most significant developments in
the decade has been the reduction in influence of
the integrated major oil companies. Peter Baxendell
traces this process from the early sixties when many
independent companies entered the market, through
the seventies, when traditional equity concessions
were phased out, to the eventual establishment of
complete producer government control of oil
production. The deintegration of the companies has
resulted in less price stability because of the
inability any longer to cross subsidise between
different markets and activities. The majors
suffered from the development of the spot markets
and short term dealing and would be adversely
affected by a collapse in the price of oil. The
case for fiscal policies designed to stimulate
energy investments is presented.

The dominance of oil in world energy markets
warrants further analysis of changes in supply and
demand. Morris Adelman shows that on the demand
side, although reductions in energy intensities have
already been very significant, because of the slow
turnover of the capital stock the process will
continue for many years and any economic growth is
likely to accelerate this intensity decline. On the
supply side the theory of resource depletion is
incapable of explaining fully the increases which
have occurred in the price of oil. It ignores the
ability of mankind to substitute knowledge for
minerals by discovering new sources and new ways of
using existing resources. User costs and scarcity
rents are much lower than is often realised so that,
at current prices, a large surplus of oil exists.
The market power of the producers is seen in their
ability to sustain a price much higher than cost,
including user costs, and the periodic crises result
from the clumsiness of cartel operations.

One question which was tackled in the last two
papers was that of why there had been so little

development of substitutes for oil. Nuclear energy
had seemed at the beginning of the decade to be an
obvious alternative to limited mineral fuels but
public fear of its possible risks have hindered its
adoption or expansion in many countries. Professor
Wolf Häfele argues that on economic grounds only,
nuclear would be preferable to a coal alternative
for power generation. However, all forms of power
investment have been loaded with regulation costs
and restrictions on safety and environmental
grounds. Given the disadvantages of present nuclear
generating plant, Professor Hafele argues the need
for the development of new energy systems in which
impurities are extracted before combustion thereby
reducing the threat to the environment.

The other alternative to oil, namely coal, has
to a limited extent substituted for oil in the
period. But this, as Herman Franssen shows, is
restricted mainly to electric power generation,
cement, iron and steel and paper making. Although
coal retains its price advantage over oil in many
markets, it faces obstacles caused by the slow
turnover of boiler stocks, industrialists' requirements of short payback periods, inadequate
distribution systems and environmental costs.

The papers reveal a substantial measure of
agreement that the world energy market will continue
to be dominated by oil for many years and that
energy prices are likely to be subject to occasional
crises as in the recent past. Only in the very
long run is it possible to envisage any substantial
fall in relative oil prices.

ENERGY ECONOMICS AND POLICY: LESSONS FROM THE PAST

Rt. Hon. Nigel Lawson M.P.

Mr Chairman,

I shall begin with a quotation "In the light of the
post October 1973 energy crisis and, in particular,
the rocketing price of imported oil, still further
steps were needed. Accordingly we set up a
Department of Energy under a Secretary of State".
That was from the Conservative Election Manifesto of
February 1974 and I have to confess that when I
wrote those words less than 10 years ago I little
thought to find myself the inheritor of those new
fangled responsibilities, invited by so distinguished and informed a gathering as this to reflect with
the benefit of hindsight on that traumatic event and
its aftermath and to venture one or two conclusions.
     Initially the great oil crisis of 1973 was
perceived as a problem of an actual physical
shortage of energy. The Arab members of OPEC, as
you will recall, banned all oil exports to Holland
and the USA as a reprisal against those two
countries' pro Israeli stance during the October
middle-east war and they announced a 25% cutback in
production in an attempt to persuade the West to
cease its military aid to Israel and to support the
Arab cause instead. The spectre of the economies of
the western world grinding to a halt because of a
physical shortage of oil was always a manifest
nonsence and indeed, pretty soon it was clear that
the real problem was something else altogether,
namely a quadrupling of the price of oil virtually
overnight and this was certainly a serious business,
presenting the world economy with an appalling
adjustment problem. But I doubt if it would have
seemed as apocalyptic as it did at the time, and to
some extent still does, had it not been for that
fleeting first phase, of the so called oil weapon.

1

Energy Economics and Policy: Lessons from the Past

Nor was the contemporary response to the genuine adjustment problem a triumph of diagnosis and prescription. The problem was diagnosed as being far more fundamental than it really was. Recall the solemn predictions of perpetual and massive OPEC surplusses which were going not merely to exert a permanent and deflationary effect on world trade, but also permanently to destabilise the international monetary system as foot loose funds washed around from one financial centre to the next. In fact, as we know, the so called non absorbers, for the most part, learnt how to absorb soon enough and the OPEC surplus no longer exists. But in prescription the boot was on the other foot because the problem really wasn't taken seriously enough. When it came to domestic policies too many consuming countries tried to pretend that nothing had happened. In the United Kingdom we got into the most almight mess, by borrowing to sustain a standard of living we could no longer afford, and in the United States, continued regulation of oil prices seriously delayed the adaptation of their economy to the end of the era of cheap energy. I hope and believe that we have all learned from our mistakes of the 70's and the lesson is surely clear, that the market economy will always adjust to an external shock, however, severe; not of course painlessly, not of course instantly, but far more effectively than statesmen are usually capable of imagining, and the task of economic management is to assist the adjustment process, not to try and resist it.

But there remains the question of why the extraordinary quadrupling of oil prices in 1973 actually happened and there are obviously lessons to be learned from this too. I believe it was the result of the somewhat fortuitous combination of a number of separate strands. Perhaps, the most important was a general one and not specific to oil, that the 1970's were the decade of inflation. The principal origin of the world-wide outbreak of this disease lay in the inflationary financing of the Vietnam War by the United States in the 60's. Up until then worldwide monetary discipline had been more or less maintained via a fixed exchange rate system based on a non inflationary dollar. When the dollar ceased to fulfil this function, the system initially became a means by which American inflation was exported worldwide. Then after the inevitable collapse of the dollar standard many countries, including notably our own, were slow to recognise

2

the need to put in its place an alternative form of
monetary discipline. One of the many consequences
of the resulting world inflation was, inevitably, a
sharp rise in almost all commodity prices during the
early seventies. Oil was not in any way alone.
Between 1972 and 1974 the prices of lead, tin and
copper more than doubled and the price of zinc
quadrupled. But, of course, there were also special
factors, affecting oil. Unlike most other
commodities the price of oil had been falling
steadily in real terms throughout the sixties.
Partly as a result of this, the United States - the
free world's biggest oil producer and the largest
oil market - moved from being an exporter of oil to
a substantial net importer and, meanwhile, the
behaviour of the major international oil companies
was changing in response to the increased powers
assumed by the governments of producing states. The
net result of all these factors was that by the time
OPEC chose to assert itself, in the wake of the 1973
Yom Kippur war it was pushing at an open door. But,
the circumstances that enabled this to happen have
not recurred, nor is there any reason to expect them
to do so.

The second oil shock, when prices very nearly
doubled in 1979, was a very different matter from
the first. What happened then was that a relatively
small reduction in world oil output as a result of
the turmoil following the Iranian revolution led to
a sharp increase in demand as oil companies and
consuming countries, haunted by the events of 1973,
sought to build up their stocks come what may.

The outcome was an oil price well above any
sustainable equilibrium level.

And this became increasingly apparent as the
effects of the world recession, which the further
oil price increase had helped to trigger, were
super-imposed on the inevitable, albeit delayed,
world response to the original shock - the response
on the supply side being the growing development of
non-OPEC energy output in general and oil production
in particular, and, on the demand side, conservation
and substitution on an increasing scale. The
elasticity of demand for oil may be low in the short
term: in the medium term it is very significant
indeed. For example, in the early 1970s, every
increase in the national product of the OECD
countries required an equal percentage increase in
their oil consumption. For the 1980s, the ratio is
expected to be only three fifths.

OPEC's response to the changed conditions of

the 'eighties has been to try to operate, for the first time, as a classic cartel: attempting to hold the price, at least in nominal terms, by controlling and limiting supply.

But this discreet attempt to allow the real oil price to fall gently failed for two reasons in particular.

First, the sharply reduced rate of world inflation has reduced the scope for eroding the real price by this route.

Second, and perhaps even more important, the denomination of the oil price in dollars coupled with the unforseen strength of the dollar against other currencies have still further perversely sustained the real world price of oil. For example, between January 1981 and the start of OPEC's long meeting last month in London, the real sterling price of crude oil to the UK rose by nearly a fifth.

Hence OPEC's need for the first time to respond to market forces by a significant cut in the nominal price of oil.

Not that there is anything intrinsically wrong in that. However, the behaviour of the oil market since the beginning of this year suggests that expectations of a change in the nominal price of oil have a far larger effect on the stock cycle than do expectations of changes in the real price, and thus are more likely to lead to large and destabilising fluctuations in the oil price, which could have serious repercussions on, inter alia, the world banking system.

For those are the practical alternatives.

The price of oil will inevitably, as with other commodities, be determined by the market – by the balance of supply and demand. But the market's adjustment to these underlying forces can be more or less smooth, more or less erratic. We have sufficient experience of the effects of oil shocks, of the costs of economc adjustment, to recognise the desirability of doing what we can to avoid unnecessary fluctuations and to achieve equilibrium by the smoothest possible route.

Certainly, BNOC's pricing proposals have had this objective very much in mind. Not that BNOC can stop North Sea prices from going where the market takes them. But it can allow time for the market's expectations to settle down, reducing the danger of an initial over-reaction and so minimising unnecessary disruption.

And the market does indeed appear to have stabilised.

If so, it is hard to see a third oil shock of the nature of either of those we experienced in the 'seventies.

The first was a once-for-all adjustment magnified by an inflationary climate which I trust has gone for good.

The second was essentially the result of a buyers' panic from which lessons have undoubtedly been learned.

And with current OPEC output little more than half its productive capacity any increase in the demand for oil as the world emerges from recession can readily be met at current prices for a considerable time to come.

But nothing is impossible; and it clearly makes sense for us as a nation to continue to develop our indigenous energy resources across the broadest possible front wherever it is economic to do so.

Hence the important North Sea tax remissions in last month's Budget, which should ensure the development of a new generation of oilfields in the 1990s.

Hence the removal of the Gas Corporation's statutory stranglehold on gas development in the UKCS, which has led to a new lease of life for the Southern Basin in particular.

Hence the Government's policy of encouraging the development of a modern, profitable, coal industry.

And hence our commitment to a continuing nuclear power programme.

In conclusion, there are two main lessons to be drawn from the past ten years.

First, the existence of energy forecasts, including forecasts of the future price of oil, is nothing more than a demonstration of the law of supply and demand.

There is a great and understandable demand for such pseudo-knowledge of the future and there is accordingly a supply of forecasts to meet the demand - just as there is always an adequate supply of gipsies in fairground tents to meet a similar demand on a more personal basis. All this is simply part of the healthy working of a market economy, and can provide many hours of innocent pleasure to those who are interested in such matters.

But it has little to do with sensible decision-taking about the supply and use of energy, usually and misleadingly called energy policy, which has to be based on the recognition that we <u>don't</u>

know what the future holds, and which has to make sense whatever the fluctuations of the marketplace.

The second lesson of the past - and one which I have no doubt will hold good for the future - is that, at the end of the day, market forces rule.

And it follows that energy policy, to make sense, must work with the grain of market forces.

But the marketplace is no textbook model, no abstract set of equations. It is a bustling, imperfect, real world of conflict and co-operation, of risk and reward, of expectations and uncertainties.

In the successful development of North Sea oil, it has, for all its imperfections, served this country very well; and in the new phase in which we find ourselves, I am confident that it will continue to do so.

And it is within this framework that we shall be seeking to promote stability without rigidity, in our own interest as well as that of the world as a whole.

For we do hold something of a special position in the world of energy. Not the position that is attributed to us, which exaggerates in equal measure the UK's influence on the world oil scene and the world oil scene's influence on the fortunes of the UK. But because we do have a foot in both camps, as both a significant oil exporter and a major non-oil trading nation, we are uniquely placed to make a balanced assessment of the forces at work and where the western world's best interests lie.

And that, as a Government, is what we have sought to do.

THE CHANGING ENERGY MARKET:
WHAT CAN WE LEARN FROM THE LAST TEN YEARS?
Colin Robinson

1. OBJECTIVE

The main aim of this paper is to analyse some of the
principal features of the postwar energy market in
the hope that we can extract a number of lessons
which will help when we contemplate the very
uncertain energy future. Understanding the past is
a necessary (though not a sufficient) condition for
successful analysis of the future. Even though our
capacity to understand the past is so imperfect that
we invariably find it difficult to learn from
history its real lessons for the future, at least we
can hope to avoid the gross errors of those whose
"forecasts" are little more than extrapolations of
the trends they have most recently perceived.
My retrospective view is mainly in aggregative
terms, dealing with broad trends in the world energy
market; inevitably, given the great signifance of
changes in the oil market, much of the paper is
devoted to analysing those changes and their
effects. In the present uncertain state of world
energy I am happy enough to leave detailed comment
on energy prospects to subsequent speakers.
However, in the latter part of the paper I shall
venture a few remarks about general trends in the
energy market of the future[1].

2. POSTWAR ENERGY TRENDS

A summary of postwar world energy trends will remind
us of the salient features of the period before we
attempt to draw any lessons from it.

2.1 Substitution of Oil and Gas for Coal, 1950-1973

The 1950s and 1960s now appear as a relatively
untroubled period (as compared with previous or
subsequent economic history) in which economic

7

growth proceeded uninterrupted by major recessions.
Real gross product of the world as a whole increased
at an average annual compund rate of approximately 5
per cent from 1950 to the early 1970s; the annual ·
average growth rate of world "commercial" energy
consumption, as measured by the United Nations coal
equivalent statistics, was also about 5 per cent
(Table 1). Moreover, within the period of over 20
years which we are now considering the rate of
increase of energy consumption was fairly steady
with little variation from one five year period to
another.
     There were, however, fundamental changes
occurring within this steadily growing total energy
market, evidently as a consequence of movements in
relative fuel prices. As Table 1 illustrates, the
substitution of oil and gas for coal was on a very
large scale. Even though world consumption of coal
increased, its share of world commercial energy was
almost halved between 1950 and 1973 from 61 per cent
to 32 per cent (using the UN's measurement
conventions). In some of the older-established coal
producing regions such as Western Europe where
mining methods were still relatively
labour-intensive, coal industries were in absolute
decline as rising wages and other costs were
translated into rising coal prices relative to oil
prices.
     Changes in oil prices in the 1950s and 1960s
are not easy to measure with any accuracy, since
there was widespread discounting of crude oil export
prices, especially in the 1960s as oil industry
competition increased, and since there were
considerable variations from country to country in
product prices. The general trend towards falling
real oil prices is, however, quite clear. So far as
one can tell, piecing together such scattered
information as there is about discounting, the order
of magnitude of the decline in real crude oil
prices[2] was about 2.5 per cent per annum compound on
average in the 1950s and about 4.5 per cent per
annum between 1960 and the first half of 1970[3].
Since costs were also declining in real terms in
marine transportation (where larger and larger crude
carriers came into service), in refining and in
distribution, product prices in real terms tended to
fall in most countries[4]. These price signals
induced consumers, producers and associated fuel
equipment suppliers to react in the ways micro-
economic theory would predict, though with the time
delays inherent in the fuel market (see 3.4 below).

8

TABLE 1

WORLD CONSUMPTION OF COMMERCIAL ENERGY 1950 – 1973

| | 1950 | | 1973 | | Average Annual Compound Rates of Increase (%) |
	Million Tonnes Coal Equivalent	% of Total	Million Tonnes Coal Equivalent	% of Total	
SOLID FUELS	1534	61	2452	32	2.1
LIQUID FUELS	672	27	3578	46	7.5
NATURAL GAS	244	10	1517	20	8.3
HYDRO-NUCLEAR	42	2	187	2	6.7
TOTAL	2492	100	7734	100	5.0

SOURCE: UNITED NATIONS: WORLD ENERGY SUPPLIES, 1950–74

As oil prices fell relative to coal prices,
consumers had a powerful incentive to substitute oil
for coal in a widening range of uses - for example,
burning under boilers, gasification and chemical
feedstock. As a consequence, oil's share of world
commercial energy consumption rose from just over a
quarter in 1950 to over 45 per cent in 1973
(Table 1). Naturally the equipment suppliers and
the oil industry responded to the change in relative
prices by providing the increased supplies of
fuel-using equipment and fuel which were required.
Output of crude oil increased rapidly (Table 2), not
so much from the older-established areas of North
and South America as from the Middle East (where
output quadrupled between 1960 and 1973), Africa and
the centrally planned countries. High volume paid
in the oil industry, where the major companies made
profits by moving large quantities of crude oil
through their vertically integrated systems. We
shall return later (3.2 below) to the depletion
signals which oil companies observed in this period.
      In the natural gas market too, radical changes
occurred. Improvements in technology allowed
natural gas to be traded internationally (by
pipeline and in liquefied state by tanker) as well
as in the countries of production. Gas was then a
relatively cheap fuel - many would say it was
under-priced compared to other fuels - and its share
of world energy doubled from 10 per cent to 20 per
cent between 1950 and 1973 (Table 1).
      Although the pre-1973 world may now seem
remote and even irrelevant, some important lessons
can be learned from the experiences of those times.
In particular, they provide significant illustration
of how the interplay of consumers and suppliers can
produce sweeping changes in a comparatively short
period. The energy market is a long way from the
"perfect" market which some economists regard as
desirable and the process by which demand and supply
interact with price is far more complex than
economists' expositions of the price elasticity
concept often suggest. Estimated elasticities of
supply and demand with respect to price are no more
than superficial indicators of complicated processes
of technological and behavioural changes initiated
by those who are affected by price variations and
whose price expectations therefore change (see 3.4
below). Nevertheless, the evidence of the 1950s and
1960s is that relative price movements are potent
forces of change even if they take time to work.
How else can one explain the halving of coal's share

TABLE 2

WORLD OIL PRODUCTION, 1960-1973

Million tonnes

	1960	1973	Increase 1960-1973
North America	410.5	621.3	210.8
Middle East	261.8	1052.5	790.7
Latin America*	194.9	272.1	77.2
Africa	13.8	290.0	276.2
USSR, Eastern Europe & China	167.2	503.0	335.8
Others	42.4	132.8	90.4
	1090.6	2871.7	1781.1

*   including Mexico

SOURCE: BP STATISTICAL REVIEWS OF THE WORLD OIL
          INDUSTRY (ANNUAL) AND BP STATISTICAL REVIEW
          OF WORLD ENERGY 1982

of world energy (from nearly two thirds to one third) and the doubling of the combined share of oil and gas (from just over one third to two thirds) in a period of just over twenty years?

2.2   Price Changes in the 1970s and early 1980s

In the 1970s, the price messages emanating from the fuel market changed abruptly, primarily because of upheavals in the world oil market. The declining trend of real crude oil export prices reached its lowest point early in 1970; by the latter part of that year, after cuts in Libyan crude output, prices were increasing. Then in 1973-74 and again in 1979-80 there were two huge upward steps in prices. Figure 1 illustrates the movement of official sales prices (and their predecessors, posted prices) from 1950 to 1982, showing both nominal and real changes.

The massive scale of the increases can more accurately be gauged by examining market prices rather than the more stable officially quoted sales prices. For instance, the market price of Saudi Arabian light crude early in 1970 appears to have been approximately $1.25 per barrel, whereas by the latter part of 1980 it had risen to a peak of over $40 per barrel on the spot market (an increase of over thirty times). Even after deflation to real terms[2] the increase in just over ten years was about ten times. Furthermore, the market has behaved erratically. After the 1973-74 price increases, crude prices drifted downwards in real terms until late 1978 and after the second price "shock" in 1979-80 prices fell more substantially (although the impact on the oil exporters was cushioned by the appreciation of the dollar from 1980 onwards). By early 1983 the spot market price of Saudi light was around $28 per barrel (over 30 per cent below its late 1980 peak) and in March 1983 the official price was reduced to $29.

3. LESSONS FROM RECENT HISTORY

3.1   Some questions

When we contemplate the dramatic events of the last ten years and contrast them with the more stable earlier postwar period, four questions arise which require answers. First, how is it possible for any market to change so fundamentally in so short a period as did the oil market from the 1960s to the 1970s? Second, what have been the extent so far of the demand and supply responses to fuel price variations? Third, what obstacles are there to energy market adjustment away from oil and gas?

Finally, why has the oil market become so volatile
in the last five years? Unless we can offer
reasonable answers to such questions - albeit with
the advantage of hindsight - there is little chance
we can make any useful observations about the
future. Of course, even with convincing
hindsight-assisted answers, we are unlikely to
achieve more than a rather blurred vision of energy
prospects.

3.2   Why did the oil market change in the 1970s?

The first of these questions is in many ways the
most difficult. The popular simple explanation of
the crude oil price rises, which regards OPEC as the
only villain of the piece is rather unconvincing;
since OPEC was formed in 1960 there must have been
reasons why its behaviour - or, more precisely, the
behaviour of the major oil producers - changed
between the 1960s and the 1970s. Most likely OPEC's
role in the price increases is generally
exaggerated, although the Organisation serves as a
convenient scapegoat[5]. The answer to the question
of whether OPEC _per se_ has raised oil prices is
rather like the answer to that time-honoured
question asked of economics undergraduates - do
trade unions raise wages? The answer is that both
OPEC and trades unions on occasions have actively
increased the prices of the products they supply: on
other occasions they have been merely vehicles for
translating market pressures into higher prices and
wages; and on yet other occasions they have acted as
moderating forces tending to keep prices and wages
below their market clearing levels.
    By the early 1970s the members of OPEC had
gained in confidence as well as in numbers, they
were anxious to lead a Third World crusade against
"exploitation" and they were both more willing and
more able to exploit monopoly power than they had
been in the 1960s. The 1973 Arab-Israeli war also
provided an occasion for the Arab members of OPEc to
take supply-restricting action. But in the 1970s
OPEC had no formal output-sharing scheme, as a true
cartel would have done. Perceptions of the power of
OPEC or of major producers such as Saudi Arabia,
Iran and Iraq may well have been a significant
factor in generating fears of scarcity in the early
1970s but the huge price increases which occurred
would hardly have been possible had background
economic forces not been propitious.
    The key to the change in economic forces can
probably be found within the standard theory of
resource depletion as developed by economists over

the last hundred years. One can conclude from the
theory that oil producers will make decisions about
producing marginal barrels or holding them back for
the future (relative to planned output programmes)
according to whether they think oil in the ground is
likely to be more or less valuable than money
invested[6]. In more technical terms, producers will
compare expected rates of price appreciation (net of
costs) with their discount rates.

Oil companies producing in the Middle East in
the 1960s probably had rather limited time horizons,
because it is likely they anticipated partial or
complete takeover of their producing operations and
thus their discount rates were higher than they
would normally have been; my own experience
indicates that they also subscribed to the general
expectation of the time that real oil prices would
remain approximately constant or decline for many
years. The resulting coincidence of high discount
rates and low price expectations appears to have
resulted in a strong tendency to produce oil sooner
rather than later, thus holding prices down. World
oil output more than doubled between 1960 and 1970.
However, as oil output grew rapidly, fears of future
scarcity eventually emerged and price expectations
changed from the late 1960s onwards. At the same
time "host" countries with relatively long time
horizons took over producing decisions and to them
money in the bank began to look a poor proposition
compared with the apparently excellent prospects of
price appreciation if oil was left in the ground.
There can have been few better investments than a
barrel of oil left in the ground early in 1970,
extracted late in 1980 and sold at a price ten times
as high in real terms (with only a small increase in
production costs). Though ex ante no one
anticipated such huge increases, there is no doubt
that in the early 1970s expectations of big price
increases were formed – a common view was that crude
prices would double or treble by the 1980s[6]. These
altered expectations were important determinants of
the change from rapidly rising world production to
constant or modestly increasing output. Enhanced
price expectations coupled with lower discount rates
gave a powerful incentive to hold marginal barrels
of oil in the ground. Thus supply behaviour
altered. Producers' supply curves shifted to the
left along demand curves which in the short-run,
were very inelastic with respect to price and prices
therefore rose sharply. Demand curves shifted too,
because of reduced real income, but not by enough to

14

avoid price increases induced from the supply side.
I have emphasised the significance of the
underlying economic forces in the oil market of the
early 1970s rather than the power of OPEC per se.
Although one cannot readily disentangle their
separate efforts on prices, it seems highly unlikely
that, in a market with a considerable margin between
price and extraction cost, an oligopoly without an
explicit output allocation scheme can for long fix
prices. However, for the reasons we have given,
market circumstances in the early 1970s made
output-restricting behaviour appear unusually
advantageous to oil producers. Consequently, there
was sufficient community of interest for those
producers - at that time mainly concentrated in OPEC
whose members supplied over 90 per cent of world
crude oil exports in 1972[8] - to be able to exert
market power even without production quotas.

Since the mid 1970s though, it is not clear
that OPEC per se has played any significant part in
such price increases as there have been. The
Organisation certainly placed a limit on price
reductions in the period from late 1981 onwards when
spot prices fell well below official prices (see 3.5
below). But, looking back at the 1970s, we can see
that by about 1977 the oil market was already
showing signs of weakness and the influence of OPEC
seemed to be on the wane. The market changes then
occurring were masked by the price explosion of
1979-80 which had very little to do with OPEC. The
second price shock was attributable primarily to
anticipations of shortage - which revived
expectations of higher future prices - in that very
uncertain period which followed first the Iranian
revolution and then the early stages of the Gulf
War[9].

As prices rose in 1979-80 OPEC seems to have
acted as little more than a passive vehicle for
translating into "official" prices market movements
which had already occurred. Dramatic meetings in
exotic places were widely reported and the world's
Press seems to have been led to believe that the
Organisation was engineering price increases. The
truth was more mundane. OPEC meetings took place in
very confused market circumstances and were
essentially attempts to interpret, ex post, the
market's signals to crude producers about where they
could set their prices and what differentials could
reasonably be fixed in relation to light Arabian
"Marker" crude. Indeed, some OPEC members such as
Saudi Arabia tried to moderate the price increases

of 1979-80. Then from late 1981 onwards OPEC had
the novel experience of trying to cope with a
falling market, first by reductions for some of the
over-priced crudes and then by a partially-observed
production quota scheme instituted early in 1982.
    Thus, over the last seven or eight years there
is little evidence to support the popular view of
OPEC as the price-maker in the world oil market,
though of course the behaviour of individual
producing countries has had significant effects.
Not only Saudi Arabia but also Mexico and Britain[10]
have had considerable influence on world prices in
recent times.

3.3   To what extent have energy markets adjusted?

We have already commented (2.1) on the powerful
effects of the relative price changes of the 1950s
and 1960s when oil and gas were substituting for
coal. From the early 1970s to late 1980 the
direction of price movements was reversed as oil
prices rose sharply relative to other fuel prices.
The three principal results were that energy
consumption increased much more slowly than in the
earlier period, that there was virtually no growth
at all in world oil consumption (indeed it fell from
1979 onwards) and that substitution of other fuels
for oil began. Falling oil prices in the last two
years have not so far had time to show significant
effects in energy markets so I concentrate on the
market effects of the price increases.
    In this summary review, we can illustrate the
main energy market trends with some statistics
derived from British Petroleum's Statistical Reviews
of the World Oil Industry and of World Energy.
Table 3 compares world energy consumption trends
(expressed in oil equivalent, not coal equivalent as
in Table 1) in the nine years before 1973 with
trends in the 1973-82 period. The drop in the rate
of growth of world energy consumption is from over 5
per cent per annum to about 1.5 per cent per annum;
oil consumption growth was reduced much more, from
nearly 8 per cent per annum to practically zero; the
rate of growth of natural gas was cut by two thirds;
but coal consumption increased faster than in the
earlier period. Nuclear power growth after 1973 was
somewhat less in percentage terms than in 1964-73,
though from its low base it was still rapid compared
with other fuels, and the rate of increase of hydro
power was rather lower after 1973 than before.
    In terms of market shares, the main results
were that oil's share fell considerably from 47 to

TABLE 3

WORLD PRIMARY ENERGY CONSUMPTION, 1964-82

	1964		1973		1982		AVERAGE ANNUAL COMPOUND RATES OF INCREASE %	
	M.T.O.E.	% of Total	M.T.O.E.	% of Total	M.T.O.E.	% of TOTAL	1964-73	1973-82
OIL	1420	37.8	2798	47.3	2819	41.2	7.8	0.1
SOLID FUELS	1505*	40.1	1668	28.2	2041	29.9	1.1	2.3
NATURAL GAS	604	16.1	1076	18.2	1312	19.2	6.6	2.2
NUCLEAR	4	0.1	49	0.8	217	3.2	32.1	18.0
HYDRO	221	5.9	332	5.5	446	6.5	4.5	3.3
TOTAL	3754	100.0	5923	100.0	6835	100.0	5.2	1.6

* PARTLY ESTIMATED

SOURCES: BP STATISTICAL REVIEWS OF THE WORLD OIL INDUSTRY (ANNUAL)
AND BP STATISTICAL REVIEW OF WORLD ENERGY 1982

17

41 per cent, the share of gas rose a little from 18 to 19 per cent and there were signs of a revival in coal, after many years of relative decline, as its market share increased by almost two percentage points from 28 to 30 per unit. The share of nuclear electricity in world energy rose from nearly 1 per cent in 1973 to about 3 per cent in 1981 and hydro electricity also increased its share slightly[11].

Disaggregation by region (Table 4) shows that energy consumption in the industrial countries of North America, Western Europe and Japan was slightly lower in 1982 than in 1973. Such growth as occurred in this period was in the developing countries (in many of which relatively high rates of real GNP increase continued after 1973) and in the communist world. In the same two areas, oil consumption kept on rising after 1973 (Table 5): in the developing countries, for example, it increased at an average annual compound rate of nearly 4.5 per cent per annum from 1973 to 1982 and in the communist countries the annual rate of increase was over 3.5 per cent. Since oil consumption North America, Western Europe and Japan dropped considerably over the same period, by 1982 they accounted for only 56 per cent of world oil consumption compared with nearly 70 per cent in 1973.

In a transition period containing two major price shocks, one would expect that the trends we can perceive for the period as a whole would be accentuated in its latter part. That indeed is the case. Since 1979 energy and oil consumption have fallen in the industrial countries and elsewhere they have risen more slowly than in 1973-79. Table 6 shows that world energy consumption has fallen slightly in each of the last three years and that oil consumption dropped by 3 to 4 per cent in each year. Although the size of the oil consumption decline is almost certainly exaggerated - because in 1979 supply uncertainties and rising prices must have induced final consumers to increase their stocks whereas subsequently those stocks must have been reduced - the changed trend of oil consumption is most significant. After about a hundred years of virtually uninterrupted rapid growth, world oil consumption has recently shown clear signs of decline.

3.4   Obstacles to Market Adjustments

In this brief survey of the principal features of the recent past we need to consider what constraints there are on the energy market adjustment process.

TABLE 4

WORLD PRIMARY ENERGY CONSUMPTION BY REGION

	1973 M.T.O.E.	1982 M.T.O.E.	Average Annual Compound Rate of Change 1973–82 %
North America	2014	1936	−0.4
Western Europe	1241	1217	−0.2
Japan	348	340	−0.3
Australasia	67	90	+3.3
USSR, Eastern Europe & China	1594	2220	+3.7
Latin America, Africa, Middle East, South & S.E. Asia	659	1032	+5.1
TOTAL	5923	6835	+1.6

SOURCE: BP STATISTICAL REVIEW OF WORLD ENERGY 1982

TABLE 5

WORLD OIL CONSUMPTION BY REGION

	1973 Million Tonnes	1982 Million Tonnes	Average Annual Compound Rate of Change 1973-82 %
North America	902	776	-1.7
Western Europe	749	601	-2.4
Japan	269	207	-2.9
Australasia	35	36	+0.3
USSR, Eastern Europe & China	455	632	+3.7
Latin America, Africa, Middle East, South & S.E. Asia	388	567	+4.3
TOTAL	2798	2819	+0.1

SOURCE: BP STATISTICAL REVIEW OF WORLD ENERGY 1982

TABLE 6

WORLD PRIMARY ENERGY CONSUMPTION BY FUEL

MILLION TONNES OIL EQUIVALENT

	1979	1980	1981	1982
OIL	3125	3002	2906	2819
SOLID FUELS	1976	2007	2003	2041
NATURAL GAS	1278	1306	1327	1312
NUCLEAR	153	169	198	217
HYDRO	412	420	428	446
TOTAL	6944	6904	6862	6835

SOURCES: BP STATISTICAL REVIEWS OF THE WORLD OIL INDUSTRY
(ANNUAL)

AND BP STATISTICAL REVIEW OF WORLD ENERGY 1982

The Changing Energy Market

Although energy and oil consumption have clearly
reacted to price increases it seems that, as one
would expect on theoretical grounds, demand is
inelastic in the short run with respect to price[12].
On the supply side of the market there is little
sign of any significant expansion of non-oil sources
of energy, though Table 3 indicated a limited
revival in coal and some growth in nuclear power.
It is worth examining some of the apparent reasons
why there have not so far been greater demand and
supply responses to the massive changes in oil and
other fuel prices in the 1970-80 period because such
an examination should help us determine what
obstacles to adjustment there may be in the future.
     First, consumer prices have risen much less
than the export prices of crude oil on which
attention tends to be concentrated. Depressed
tanker freight rates and comparatively small
increases in oil refining, distribution and
marketing costs have damped down increases in pre-
tax consumer prices as compared with f.o.b. crude
prices. Government policies have generally
reinforced this dampening effort and have thereby
tended to hinder adjustment. Some countries have
deliberately tried to hold their fuel prices below
the world level and almost everywhere taxes on oil
products have been increased relatively little. The
price signals received by consumers of oil and
competing fuels have therefore been much less strong
than an examination of crude oil export price
increases would suggest.
     Second, on the supply side, the strength of
the signals reaching fuel producers has also been
reduced by the tendency of governments to regard
increases in producer profits consequential on
rising oil prices as a windfall which can be taxed
away without affecting supply behaviour. There is,
in fact, good reason to believe that the prospect of
comparatively high profits for a period is a
necessary incentive if actual and potential fuel
producers are to make the investments required for a
powerful supply response. If governments impose
high rates of profits taxation and lead producers to
expect that any future price increases will bring
still higher tax rates - as the British and
Norwegian governments have led North Sea producers
to believe - they or their successors will reap the
consequences in terms of a reduced willingness to
invest.
     The third and fourth constraints on adjustment
are inherent time lags in demand and supply

responses which make fuel markets much more complex
than the instantaneously adjusting systems of the
elementary economics textbooks. Because of the
complementarity between fuels and fuel-consuming
appliances, demand reactions to price, income or
other variations are necessarily slow. When for
example, fuel prices change relative to prices in
general and relative to each other, it takes time
for consumers' price expectations to change and for
their behaviour to alter so that they make the
investment decisions which will begin to adjust
their equipment stocks to those desired on the basis
of new prices and new price expectations. The
characteristics of fuel-using equipment also respond
over a period of time to price changes, as we have
seen car design change in recent years now that
increasing fuel prices have made economy a selling
point.

On the supply side there are also time lags
which are inevitable given the long periods it takes
to bring into operation such major new facilities as
oilfields, coal mines, power stations and
refineries.

Such inherent delays on the supply side
should, however, be distinguished from another form
of delay which represents our fifth constraint and
which has become very important in recent years as
concern about environmental damage has grown. It is
irrelevant to this paper whether such concern is
over-done or not. New energy supply facilities tend
to be obtrusive and to carry potential pollution
hazards. Thus we must anticipate that objectors
will delay their introduction, whether they are
offshore oilfields in the United States, hydro
schemes in Norway or coal mines in Britain.

The biggest problem of all with public
opinion, however, relates to nuclear power. It is a
curious paradox that the replacement energy source
which was most advanced by the early 1970s happened
to be the one which turned out to arouse most public
anxiety and opposition. Nuclear power programmes
are seriously delayed almost everywhere in the world
(except for France and a few small countries where
the delays are so far minor).

Although nuclear power's share of world energy
has increased quite rapidly since 1973 (Table 3) and
it now provides about 9 per cent of world
electricity[13], its growth has been much less than
was generally anticipated in the early 1970s. In
1973 projections of OECD nuclear capacity were for
over 500 GW (net) in 1985 and over 1000 GW in 1990.

By 1982 the 1985 projections were reduced to 220-240 GW and the 1990 projections were 330-370 GW[14]. Even these much lower projections seem unlikely to be achieved. To some extent, delays in nuclear pro- grammes are attributable to the technical problems which afflict new technologies and to the labour relations difficulties common on large construction sites. But the main factor now seems to be failure to convince an articulate section of the public that a big expansion of nuclear capacity is needed, particularly since electricity consumption is rising only slowly in the industrial countries.

3.5   A fluctuating market

So far this paper has concentrated on the oil price increases in the 1970-80 period and their effects, mentioning only in passing the subsequent drop in prices. We must now turn to the most recent events in the world oil market.

Figure 2 provides the statistical background, illustrating in particular the considerable fluctuations in spot prices since the beginning of 1979. During 1979 official prices (measured in the Figure by Saudi light crude) were increasing in the aftermath of the Iranian Revolution, though lagging well behind spot prices; in October 1979, for example, the spot price ($38) was more than double the official price ($18). Spot prices remained well above official prices throughout 1980, reaching a peak of just over $41 in November of that year. The official price, however, did not peak until October 1981; by that time the spot price had subsided and there was a period in the second half of 1981 when spot and official prices were approximately equal in the range $32-34 per barrel. As the market weakened in 1982, spot prices fluctuated a good deal but generally remained well below official prices (which were by then being discounted by a number of countries) until in February and March 1983 the official price was reduced to about the then spot market level.

Recent experience seems to suggest yet another change in the nature of the world oil market. In the 1950s and 1960s, price trends were fairly smooth (Figure 1); then in the early 1970s there was the first of the two massive upward steps discussed earlier in this paper. After the first oil price shock the market turned temporarily weak until surplus gave way to scarcity in 1979-80 and to renewed surplus from late 1981 onwards.

This tendency to swing from surplus to

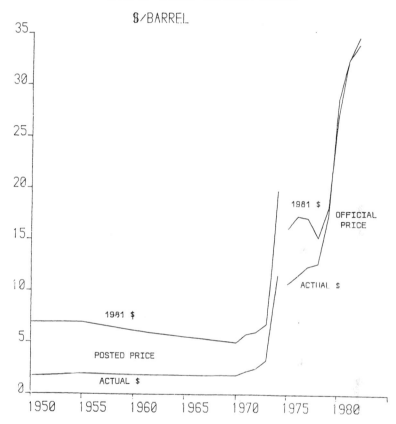

FIGURE 1

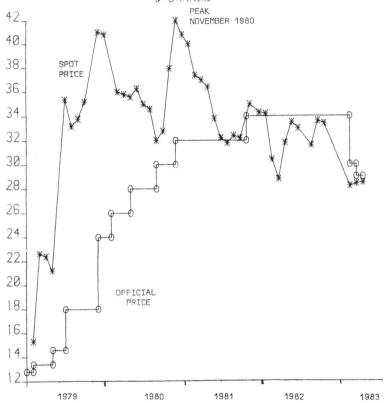

EXPORT PRICE OF SAUDI LIGHT CRUDE OIL
$/BARREL

PEAK
NOVEMBER 1980

SPOT
PRICE

OFFICIAL
PRICE

1979   1980   1981   1982   1983

SOURCES: OPEC Bulletin Petroleum Economist Middle East Economic Survey

FIGURE 2

scarcity and back again, with erratic variations en
route, suggests that the oil market is much more
prone than it used to be to an inventory cycle of
the sort which causes fluctuations both in the
economy as a whole and in commodity markets in
particular. Many years ago economists demonstrated
how a model which combines the multiplier and the
accelerator can, on plausible assumptions about the
values of the coefficients, generate cyclical
movements in the economy[15]. Using similar reasoning
to consider the case of the oil market, we might
describe the cycle and its trigger mechanisms in the
following stylised fashion.

Let us postulate as a starting point some
event in the oil market which causes expectations of
scarcity and thus of rising prices. Given the
experience of the 1970s such an event - which might
be a revolution in an oil-producing country, a spurt
in economic growth, or possibly just a very cold
winter - will probably cause a scramble for
supplies. Added to the demand from consumers will
therefore be a demand for oil for inventory, so that
demand on the oil producers increases sharply for a
period as storage tanks are filled and tankers steam
more slowly. Supply is also likely to be restricted
as producers see some advantage from holding off the
market to take advantage of higher expected prices.

The rate of addition to inventory must, however, decline as physical limits of stock holding
come near. Moreover, rising prices, because of the
short run inelasticity of oil demand, tend to
produce rising revenues for the oil producing
countries a relatively high proportion of which
(again in the short run) is saved and which
depresses real GNP in the industrial world. The
governments of industrial countries may accentuate
the recession by deflationary action designed to
offset incipient balance of payments deficits caused
by the increasing cost of oil imports. Demand on
the producers both for inventory and for final
consumption will therefore at some stage begin to
fall and price expectations will also tend to be
depressed. Stock holders will then compare the poor
prospects for price appreciation with the interest
and other costs of holding inventories, decide to
reduce those inventories and thereby depress demand
on the producers below the level of final consumption. By this time, producers will be concerned
about declining revenues as prices drop and will
start fighting for market share, thus depressing
prices still more. After a while, however, there

will probably be an increase in oil consumption in
response to the price fall and to rising real
incomes, price expectations will reverse,
inventories will start to be re-built and there
will be a new upswing in the oil inventory cycle.

Of course, in reality the inventory cycle is
less regular and mechanical than suggested and it is
clearly subject to shocks which may cause an upswing
or a down swing to be accentuated and prolonged.
There have been two very large shocks on the upside
in the last ten years. A shock in the opposite
direction has recently occurred in the form of a
drop in prices; so far the downward shock has been
proportionately much smaller although in a market
with such a large margin between price and
extraction cost is is always possible that further
discounting will generate a price war. But whatever
happens to oil prices in 1983 the important point is
that the oil market seems to have become much less
stable than it used to be. How can we explain that
change?

The most likely explanation relates to the
important alterations in the structure of the
international oil industry which have occurred in
recent times. We have already referred to the
changes in property rights as control of production
decisions passed from the oil companies to the
producing countries from the early 1970s onwards.

It is estimated that in 1973 the seven major
oil companies had 30 million barrels a day (b/d) of
equity or preferential crude oil[16]. This was
substantially in excess of their refinery
requirements and so almost 7 million b/d was sold to
third-party customers. Access to equity oil almost
halved between 1973 and 1978 but because the majors
were granted preferential terms in several countries
their supplies of crude oil were not dramatically
reduced in this period although third party sales
had to be reduced to less than 4 million b/d. A
more serious disruption of supply channels occurred
from late 1978 onwards. We have already referred
briefly to the events following the Iranian
Revolution, the main facts about which appear to be
as follows. Early in 1979, Iranian output fell from
6 million b/d to about 1 million b/d; although there
was a temporary recovery later in the year, by mid
1980 Iranian production was down again to 1 million
b/d and Iraqi output was reduced about 3 million b/d
late in 1980 by the onset of the Gulf War.

Ex post, we can see that other oil producers
increased output and that supply was not seriously

out of balance with (reduced) demand, but that is
not how the situation was perceived ex ante.
Indeed, the period was one of tremendous disruption
and uncertainty. Companies previously heavily
reliant on Iranian crude were forced to cancel most
of their third party sales, and both the major and
smaller oil companies had little option but to move
into the spot market and to make such deals with
producing country governments as they could. Hence
the surge in spot market prices which Figure 2
shows.

It is not relevant to this paper whether the
reduced role of the international majors in world
oil is good or bad. I simply wish to observe that
in the year before 1973 when the seven majors
disposed of around three quarters of internationally
traded crude oil such market fluctuations as there
were tended to be smoothed out within their large
vertically integrated systems. Now that there has
been a substantial de-integration of the market such
smoothing no longer occurs. The world oil market
fluctuates much more freely in the way other
commodity markets do, with shocks (major or minor)
amplified by the effects of price expectations on
inventory holdings. The state of price expectations
is thus extremely important both in the long run and
in the short run. Given the distribution of
property rights, it is a prime determinant of the
rate at which oil stocks under the ground come to
market (3.2 above) and of the rate of addition to or
depletion of oil stocks above the ground.

4. LESSONS FOR THE FUTURE

I want in the final part of this paper to see
whether from the answers we have suggested we can
distil some thoughts which will help to reduce the
uncertainty of the energy future. "Forecasts" in
the traditional sense are inappropriate and probably
misleading, but we may be able to eliminate some
possibilities which seem very unlikely and to pick
out some trends which seem probable. I will deal
only with a few broad issues to avoid trespassing on
the ground covered by later speakers.

4.1 A fluctuating oil market

It seems probable that the oil market will in the
1980s be subject to fluctuations similar to those we
have seen in recent years. I tried to explain these
fluctuations (3.5 above) in terms of an inventory
cycle, which had been revealed and accentuated by
changes in the structure of the world oil industry.

Oil producing countries are now trying to integrate forwards - by establishing export refineries and petrochemical plants and by buying into downstream activities in the consuming countries - so it is possible there will one day be a re-integration of the world oil-industry. However, it is likely to be many years before an effective new smoothing mechanism appears.

That does not, however, mean that prices will plunge out of control to a level determined by extraction costs and rise explosively to a maximum set by the price of substitutes. Endogenous changes in price expectations (see 3.5 above) should be sufficient in themselves to narrow the amplitude of the fluctuations. Moreover, students of the economy-wide inventory cycle will know that exogenous floors and ceilings may help to moderate cyclical variations. In the oil market, the effective floor may be set by some output-restricting group (OPEC or its successor). Despite all OPEC's problems, it has contrived to set some kind of floor (though not as high as it wanted) under the oil market in the last two years. The prime role of the output restricting group in the future may well be to insert a ratchet at the lower end of the cycle.

What will determine the market ceiling (other than endogenous forces, such as the reversal of expectations) is less easy to see. In the short run, other fuels do not place very effective limits on oil prices because of the time lags explained in 3.4 above. Perhaps OPEC or its successor will be a price-moderator in the upswing as well as in the down swing, given the desire of oil producers to avoid a repetition of the strains and tensions they suffered in the aftermath of the massive price increases of 1979-80. Otherwise, macro economic forces may set a ceiling since really big oil price increases tend to slow rates of economic growth and, in the ultimate, help tip the world into recession, thus reducing oil demand and bringing down prices.

4.2   Market forces and energy

Some of the more nonsensical beliefs held a few years ago about the scale of the energy "crisis" have now disappeared. Fortunately, we no longer hear much about the "energy gap", nor is there so much talk of the need for massive government-financed crash programmes - whether synthetic fuel programmes in the United States or protection for indigenous coal industries in Western Europe - to adjust to the energy problems of the world which are supposedly too much for markets to handle. Moreover, for obvious reasons, no one is any longer telling us that the world will need 35 million b/d of OPEC oil in the mid 1980s. A great deal of adjustment has taken place since 1973 in world energy markets, mainly on the demand side and almost entirely as a consequence of market forces; government action has done little to help and has indeed been a hindrance to the extent that fuel prices have been held down and suppliers' profits have been over-taxed.

Nevertheless, one must recognise that in a complex market such as energy "natural forces" are imperfect and slow-moving (3.4 above).Although there is no doubt some carry-over effect still to come from the big fuel price increases of the 1970s, we must expect market adjustments to continue to be slow in the future, especially if we are right in thinking that there will be substantial fluctuations in the oil market; the signals reaching consumers and producers from that market will be less clear than in the 1970s and so expectations will be less well-formed.

It is obvious that oil substitution is proceeding rather slowly. Postponements or cancellations of coal liquefaction, coal gasification, oil from shale and oil from tar sands schemes have recently been announced because of escalating costs, uncertainty about prices and doubts about government policies. A few years ago it seemed that such fuel sources might set an upper limit on oil prices, but the "upper limit" seems to be continually rising. Most of these oil substitution projects are suffering from technical difficulties. Moreover, they are themselves energy-intensive so that oil price rises are a mixed blessing, causing cost increases as well as rising prospective realisations. Nor is nuclear power a very effective competitor for oil in many countries because it has failed to win public acceptability (3.4 above); once plants under construction are (belatedly) completed there may well be a period in which little new nuclear capacity is commissioned. Since nuclear programmes were significantly delayed in an era of sharply rising fossil fuel prices and fears of supply shortages, it is difficult to believe that the environment of the late 1980s and early 1990s will be more conducive to nuclear expansion. Renewable energy sources, though possibly very important next century, are unlikely

to provide large amounts of energy before then.
Only strip-mined coal from low-cost regions (such as
Australia, the United States and South Africa) and
to a lesser extent natural gas seem capable of
displacing substantial amounts of oil from the
market. However, development of the mines, ports,
bulk carriers and inland transport facilities needed
for a flourishing international steam coal trade
will take some years yet, to say nothing of the
period necessary for consumers to convert to coal.
     The implication seems to be that the world's
move away from oil will be a very long-run and
gradual process, marked by erratic changes and
temporary upsurges in oil consumption. It seems
quite likely that in the 1980s, and perhaps in the
1990s, although oil's share of world energy will
fall, the absolute level of consumption will rise
because increasing consumption in the developing
world more than offsets a slow decline in the
industrial countries. It would not be surprising if
by the end of the century annual world oil
consumption, which is just 3 billion tonnes now, had
risen to 3 1/2 to 3 3/4 billion tonnes. That is, of
course, a far cry from the forecasts of 9 billion
tonnes which were made in the early 1970s.

4.3   Crude oil prices

Quite recently expectations about fuel prices in
general and oil prices in particular have changed.
That change is significant because, as we have seen
(3.2 above), given the distribution of property
rights, expectations are crucial in determining the
rate at which fuel stocks come to market. Since
1973 I have argued that rising real oil prices would
be a temporary phenomenon and that changing price
expectations and a rising price elasticity of demand
for oil would within the foeseeable future turn the
energy market from scarcity to surplus[17]. Thus I
have some self-interest motive for arguing now that
the turn in the market has arrived!
     However, I am doubtful whether we have yet
reached that stage. Many of the statements being
made in the Spring of 1983 about oil prices, contain
large elements of wishful thinking and a desire to
talk prices down. It is all very well to revel in
what is perceived to be the discomfiture of OPEC,
but one of the lessons we should draw from the past
is that the power of that Organisation to fix prices
has been much exaggerated. We should be less
concerned with whether the members of OPEC will
maintain their present imperfect union than with

more fundamental economic forces which will affect
the distribution of property rights and influence
price expectations.

It is true that a repetition of the huge real
price increases of the 1970s seems highly
improbable. Those increases should probably be
regarded as a sharp once-for-all upward shift which
the world economy is still attempting to digest by
reducing its oil intensity. After the big upward
steps in prices in the 1970s and given the development of substitutes for oil (albeit slowly), we can
probably assume that oil consumption is more elastic
with respect to price than it was at the price
levels of the early 1970s. Not only has the demand
curve shifted leftwards because of income and other
changes, it has also flattened. To that extent,
consumers are more resistant to price increases than
they were. Moreover, the well-formed expectations
of future scarcity and sharply rising oil prices
which existed in the 1970s - and were themselves
extremely important elements in the price increases
which occurred - have disappeared, at least
temporarily. Producers are no longer so sure that
the trend of real oil prices will be firmly upwards
so that the net present value of oil produced in the
future will be greater than the value of oil
extracted now. Output-restricting behaviour is not
the one-way bet it appeared in the 1970s, especially
to countries with high absorptive capacity anxious
to finance economic development plans. Another
limit on crude prices may be set by the forward
integration of the oil producing countries (4.1
above) which will introduce more competition into
downstream activities and, by its effect on product
prices, may constrain crude oil prices also.

Nevertheless, there appear to be several
factors which, on average, could keep oil price
expectations pointing upwards for some years yet and
could lead to some increase in real oil prices in a
fluctuating market. The first is the probable
tendency of world oil consumption to increase a
little during the halting process of substituting
other fuels for oil. Another factor may be the
rising cost of extraction from the more remote
regions (such as the Canadian Arctic) into which
exploration has moved. In a sense, such cost
increases are "artificial" since there is probably a
good deal of lower cost oil still to be found in the
Middle East. Nevertheless, political events have
moved the oil companies into high-cost areas of
production and, unless one foresees a significant

revival of exploration and development in the OPEC countries, the consequence is likely to be long run upward cost pressure on prices.

The third factor - and perhaps the biggest imponderable on the supply side - is the possibility of supply interruptions. Serious disturbance from such interruptions may seem a remote possibility at a time when the world is awash with oil, but one of the lessons we should have learned from recent times is the speed at which the market can change from surplus to scarcity. There are already many sources of tension and unresolved conflict in the oil producing world; others may well arise when oil revenues are depressed and growth aspirations are disappointed, as is happening now and will happen again if the market fluctuates in the way we have suggested. It seems likely that at least throughout the 1980s the oil market will remain sensitive, finely-balanced and very prone to supply uncertainties. It will, in other words, remain in what Shell has aptly termed "the region of vulnerability"[18] and there may well be further price increases on account of actual or perceived periods of scarcity.

Eventually there will come a time when the trend of real oil prices turns downwards. In the long run oil prices cannot continue rising relative to the price level in general - if they were doing so the supply and demand shifts would be so large and the behavioural changes so great that the real price rise would cease. But the time lags in the system are sufficiently lengthy that the long run in this context may mean the mid or late 1990s rather than the next ten years.

4.4   A general view

In general, it seems to me we are now well into the period of transition in which the world reduces its dependence on oil. But though we are travelling hopefully, we have not yet arrived; indeed, the destination may still be quite distant. We have seen that energy markets have great capacity to adjust even to very large changes in relative prices, but they do so slowly. Furthermore, the side effects in terms of structural change,unemployment and disappointed growth aspirations are painful. Further painful change seems likely in the future.

We must also be prepared for a continuation of erratic variations in energy markets which will make it even more difficult to perceive the future

direction of events than it was in the 1970s.
Energy market changes are now well above the
"awareness threshold" of public and politicians so
we must expect the media to seize on those changes
and for there to be calls for political action in
response to the latest fluctuation. No doubt there
will be plenty of pundits who will be prepared to
project the latest direction of change into the far
distant future and draw sweeping conclusions about
energy prospects. In the 1970s they thought oil
prices would rise forever; now they evidently
believe that prices will be indefinitely depressed.
One day another decisive change in fuel price trends
will, no doubt, occur and we shall for a time enjoy
a further period of "surplus" in the energy market.
But I suspect it is premature to believe the happy
day has already arrived.

NOTES AND REFERENCES

1.   See also Colin Robinson, World Energy
Prospects in the 1980s and 1990s Zeitschwift fur
Energie Wirtschaft (forthcoming).
2.   Measured by Saudi Arabian light crude oil
(34° API) deflated by the United Nations dollar
index of unit values of world exports of
manufactured goods.
3.   More details are given in Colin Robinson,
The Future of Crude Oil Prices, Surrey Energy
Economics Discussion Paper No. 12, November 1982,
which discusses a number of the issues raised in
this paper.
4.   OECD Oil - The Present Situation and
Future Prospects, 1973, Chapter X illustrates trends
in oil product prices in the 1960s and suggests some
reasons for the decline in prices.
5.   Colin Robinson, Ahmed El-Mokadem and Paul
Stevens, The Future of OPEC. Surrey Energy
Economics Discussion Paper No. 7, August 1981.
6.   A good brief explanation is in R.M.
Solow, The Economics of Resources or the Resources
of Economics, American Economic Review, May 1974.
7.   An influential article was James Akins,
The Oil Crisis - this time the wolf is here,
Foreign Affairs, April 1973.
8.   Colin Robinson, The Energy 'Crisis' and
British Coal, Institute of Economic Affairs, 1974,
page 35.
9.   A very revealing analysis of the second
oil price shock is in M.S. Robinson, The Crude Oil

Price Spiral of 1978-80, Shell International
Petroleum Co. Ltd., February 1982.
          10. For an explanation of the likely
stabilising effects on prices of rising non-OPEC
output in the late 1970s and early 1980s see Colin
Robinson and Jon Morgan, North Sea Oil in the
Future, Macmillan, 1978, chapter 1.
          11. The market shares for hydro and nuclear
electricity are not comparable as between Table 1
and Table 3 because the UN and BP statistics are
different conversion factors.
          12. For a review of econometric results see
David Hawdon and Mark Tomlinson, Energy Demand
Models in the USA and UK, Surrey Energy Economics
Discussion Papers No. 8, January 1982.
          13. UK Atomic Energy Authority, Atom,
November 1982, p.241.
          14. See Nuclear Energy Agency, Nuclear Energy
and its Fuel Cycle: Prospects to 2025, OECD, 1982,
Table IV.5. The International Energy Agency's World
Energy Outlook, 1982, (pp. 331 et seq.) also
discusses the reduction in nuclear power plans.
          15. An early exposition of such a model
(using a second order difference equation) is P.A.
Samuelson, Interactions between the Multiplier
Analysis and the Principle of Acceleration, Review
of Economic Statistics, May 1939.
          16. The statistics in this section are mainly
derived from J.H, Mohnfeld, Implications of
Structural Change, The Petroleum Economist, July
1982, pp. 269-272. See also M. Robinson, op. cit.
and John Mitchell, Anatomy of an Oil Crisis,
Zeitschwift fur Energie Wirtschaft, 2/1982.
          17. See, for example, Britain and the World
Energy Crisis: The Outlook for the 1970s, Morrell
Publications, January 1974, especially pp. 14-17,
and Energy Depletion and the Economics of OPEC,
op.cit. especially pp. 23-24.
          18. J.M. Raisman, Oil-World Supplies and
North Sea Development, address to "ANSWER"
Conference, Brussels, March 1982.

OPEC AND THE WORLD OIL MARKET IN THE 1980s

Edith Penrose

After a long and extremely arduous series of
meetings, beginning with three weeks of negotiations
and ending with twelve days of nearly non-stop
discussions that nearly exhausted their energies,
the OPEC ministers reached agreement in the second
week of March 1983 on both the allocation of
production and the price of crude oil. An overall
ceiling of 17.4 million barrels a day for 1983 was
established with the allocation among all but one of
the OPEC members of individual production quotas;
Saudi Arabai accepted, in the role of "swing"
producer, to supply the balance within the overall
ceiling between market requirements and the quotas
of other members. On the assumption that market
demand would permit 17.4m b/d, the implicit Saudi
"quota" would be about 4m b/d. All members agreed
to accept a price for the "marker" cruder of $29/b,
a reduction of $5/b from the earlier price, with
differentials for crudes from different sources and
of different qualities to be the same as those
agreed in March 1982 (with the exception of a
"temporary" price for Nigeria to permit it to recoup
some of its losses in competition with North Sea
crude). Members accepted a number of provisions
intended to ensure that neither direct nor indirect
discounting off the officially agreed prices would
occur.

The prices and quotas agreed contain a number
of anomalies which were presumably accepted as
absolutely necessary to maintain OPEC's control of
the market and to prevent competitive price cutting
to the detriment of all. The immediate result was a
firming of the market and confidently expressed
expectations that the price set would hold until the
end of 1983, at which time it was insisted (with
equal confidence) that the oil market would have

regained its strength with the onset of industrial recovery. In addition, the ending of uncertainty over prices was expected to halt, or at least significantly to slow down, the de-stocking that had been a crucial factor in the decline of demand for OPEC oil, which had fallen to less than half of world demand by the middle of 1982. In the months following the agreement the gamble seemed to have succeeded. Few denied that the future was still highly uncertain, although in the middle of April Sheik Yamani went so far as to raise his estimate of the chances of success from his initial "50-50" to "100% certain".

There can be no question that OPEC played a central role in bringing order into oil markets. But the role was a dual one, consisting on the one hand of successful negotiations to reconcile the deep conflicts of interest among OPEC members (a remarkable achievement in itself) and, on the other, of successful diplomatic efforts to secure the willingness of some important non-OPEC producers to go along with OPEC policy, explicitly from Mexico and implicitly from Britain. In announcing its own price reductions in the week following the OPEC agreement, BNOC remained within the OPEC "guide-lines" and refrained from seeking a price advantage vis-a-vis Nigeria, which all knew would have been an open challenge to Nigeria to respond, and would thus have been an open challenge to Nigeria to respond, and would thus have gone far to undermine the new OPEC policies. This implicit acceptance of OPEC's objectives was later more or less explicitly confirmed by the British Secretary of State for Energy, who publicly asserted Britain's desire to promote "stability without rigidity" in world oil markets in the interest both of Britain and the world as a whole.

It is evident that OPEC can no longer play the more-or-less lone hand it has played in the past, but it is likely to retain a central position on the world scene and perhaps a decisive one. Before considering the prospects of this, let us first clarify what we mean by "OPEC". The acronym is used very loosely and in different senses. Sometimes it is used to refer to the formal body or <u>organisation</u> of the petroleum exporting countries - its original meaning; sometimes it is used in a context that refers only to the <u>Arab</u> oil exporters (as when "OPEC" is held responsible for the embargo during the 1973 Arab-Israeli war); sometimes it is used to mean only the Gulf producers, including Iran; and at

times only Saudi Arabia. (It has even been said
that Saudi Arabi is OPEC). Even when the
organisation is clearly the subject of discussion it
is often not evident whether OPEC is looked at as a
loose oligopoly of a few major producers surrounded
by a fringe of lesser fry, or as a cartel of
thirteen countries continually functioning under the
incipient strains of defections eventually ensuring
its "collapse". It may seem to some that to raise
this question is to quibble, but in my view it is
important for this discussion because "OPEC" in its
different forms will have a different existence and
role depending on the form assumed.

As an organisation OPEC will continue to be
useful to members largely as a centre for the
exchange of information through conferences,
seminars, etc., "technical assistance" of all kinds,
ranging from advice on contracts and other issues
to data collection, and public relations
activities. But its role in oil markets as an
organisation has, in my opinion, been very much
exaggerated. At the several important turning
points in OPEC's history, events have been dominated
by one (or a very few) major producing countries
who were in a strategic position to initiate change
or to take control; the organisation as such was
only rarely of decisive importance, except, perhaps,
for the psychological impact its very existence has
had both on the oil exporting countries and
outsiders. The importance of this impact should not
be underestimated, but it is not usually perceived
as the important source of OPEC's power. It also
served an important function as an agency around
which its members could rally.

To illustrate, Colonel Ghadaffi's success in
1970 owed little or nothing to OPEC, and it was the
Gulf producers, not OPEC , who moderated the
leapfrogging process thereafter. Saudi Arabia,
together with an ally or two, was primarily
responsible for the Teheran agreement and the
subsequent framework for participation agreements -
although, to be sure, the threat of concerted OPEC
action was, at one point at least, decisive. Nor
was OPEC as an organisation responsible for the
price rise following the war in 1973, although
naturally no member cared to reject the opportunity
created. But would not a similar chain of events
have been likely even in the absence of a formal
organisation?

In fact, OPEC's short period of 'dominance' in
the 1970s was marked as much by discord as by

agreement; and given the political configuration of
the Arab world it seems likely that even without
OPEC the Gulf producers would have used their
productive power in much the same way as they did
within it. But be that as it may, the strains in
OPEC between those members wanting higher prices and
output controls and those preferring lower prices
with output determined by appropriate price
differentials have been very great. Indeed, it
could not have been otherwise given the distribution
of oil reserves and the natural desire of each to
protect its own interest. Now their differing
interests have been contained for the moment by
their desire to prevent a disaster that would
seriously impair the interests of all.

The question hanging in the air of course, is
how long their cohesion can last in view of the fact
that it has never been very solid. OPEC spokesmen
themselves like to argue that the large increases in
prices that have occurred during the past ten years
have not been the result of the exercise of OPEC
monopoly power so much as a response to market
forces. In other words, that the level of prices
cannot be attributed to the cohesion of an OPEC
cartel or monopoly. There is some truth in this,
but it is obviously not the whole truth since even
if we accept that price increases responded to the
market, price decreases - the appropriate response
in a falling market - were clearly inhibited by OPEC
policy. On the other hand, the appropriate cartel
response, the imposition of output quotas, was not
taken until the really serious crisis approached.

In general, as an effective price and output
cartel the organisation of petroleum exporting
countries cannot, and presumably would not try to,
claim great success. At the same time, however, the
OPEC countries have been very effective in raising
the prices of their crude oil. They came close to
catastrophe, partly because, with an inadequaste
appreciation of the limits to their power, they went
too far. This in turn reflected an optimistic
appraisal of trends in the international economy (in
which they were not alone!) together with an
inadequate understanding of the oil market as it
changed under the impact of high prices,
uncertainty, and consumers' apprehensions over
security of supply. It might be noted in passing
that the OPEC ministers also seem to have begun to
fear that they were not getting good enough
information from the OPEC secretariat, if one can
judge by reports that they intend to strengthen the

secretariat and try to reduce political bias in its analysis and advice.

The apparent inconsistency between the lack of cohesiveness within the organisation and the relative success of the OPEC countries in the market is precisely what had so often led, at the one extreme, to continued predictions that OPEC would "break-up" and, at the other, to the assertions that OPEC "is Saudi Arabia". I myself do not accept the argument frequently advanced that the increases in the price of oil are explained by a perception of impending depletion. Rather I think they were due simply to the desire and ability of the oil-exporting countries to take advantage of market conditions when they were favourable and to their success in holding their position when markets turned unfavourable. The ability to hold the line, to prevent a serious erosion of nominal prices must, I think, be attributed largely to Saudi Arabia, with some assistance from other Gulf producers.

Saudi Arabia was prepared to accept price increases with their consequent inevitable effects on Saudi output in order to try to maintain her political and economic leadership in both the Arab and the oil world. The economic ability and political willingness of Saudi Arabia to accept very wide swings in output enabled OPEC to maintain a high level of prices during the 1970s; for the same reason Saudi Arabia also succeeded in moderating the increases demanded by other OPEC members. The capacity of Saudi Arabia alone to regulate prices through control of her own output has, however, been tested on both the high and the low side and has been found to be limited. Nevertheless because of her weight as a producer, Saudi Arabia, together with a few other Gulf producers, must remain the core of OPEC.

From this analysis follows an appropriate definition of "OPEC" from the point of view of its past, as well as of its future role in world oil markets: OPEC is an organisation that provides through its existence and its decisions the means the mechanism for the implementation of oligo-polistic market and production policies by countries exporting crude oil. It was successful in this role from 1974 until 1981/82 when the loose oligopolistic framework that governed the individual behaviour of the member countries became increasingly inadequate and had to be replaced in 1983 by an explicit cartel regulation and monitoring of the prices and output of its members. But in its essence the period of

41

OPEC control relied on much the same type of
competitive restraint exercised by the few dominant
suppliers to world markets as had existed in the
past under different regimes.

When oil from the very large low-cost oil
fields of Saudi Arabia and Kuwait began to flow into
world markets in the later 1940s and early 1950s,
reasonably orderly (though by no means completely
stable) price movements were maintained largely
because of the restraint exercised by the few large
oil-producing international companies although, as
Adelman has convincingly argued, their restraint was
supported and encouraged by the barriers to crude-
oil imports erected by the United States and by the
excise taxes on fuel oil imposed by the coal-
producing countries of Europe[1]. The tacit control
by the companies lasted little more than ten years;
it began to break down in the late 1950s as oil
increasingly leaked into world markets supplied both
by the major companies themselves and by a few new
entrants. Product prices came under more and more
pressure in Europe and elsewhere as competition
intensified and boosted prices. The linch pin of
the prevailing price policies could not be maintained, even with a high degree of vertical
integration, given the tax regime in force.

After the formation of OPEC in 1960 prices
continued to drift downwards, but again in a
reasonably "orderly" fashion, the efforts of the
companies now being helped by the success of OPEC in
setting a floor to tax prices. This era of "price
management", too, lasted roughly a decade. By 1970
both "official" and spot prices had reached the
lowest point since 1960[2]. The "OPEC/Company" system
broke up when Colonel Ghadaffi took advantage of
very favourable circumstances to make a determined
onslaught on the price and tax arrangements that had
prevailed since the early 1950s. The transition
from company to producer government control was
accelerated by the 1973 Arab-Israeli war. Although
the "OPEC system" from 1974 was more formally
organized it, like that of the companies earlier,
depended fundamentally on oil producers being
willing to restrain competition among themselves.
It was reasonably successful until the beginning of
the 1980s.

This interpretation of the three phases of
"price management", broad brush and extremely
simplified as it is, does not, in my view, present a
misleading view of the behaviour that produced the
balance between the forces of competition and the

mixture of formal and informal attempts to regulate
the course of oil prices for the past three
decades. The question now is whether with a very
different market structure, and a very different
configuration of players both geographically and
politically, the type of behaviour that effectively
maintained reasonable order in oil markets will
again emerge. I suggest that it may well do so.

The chief difference between the past and the
situation today, which will probably be accentuated
in the future, is, of course, the much larger number
of oil-producing and oil-exporting countries whose
policies will have to be coherent with those of OPEC
if it is to play a central role in world markets.
The OPEC producers as well as the great companies
are in a much weaker position than they have been,
while the gap is even larger than before between the
oil and the cost, not only of developing and
producing known reserves but probably also of
finding and establishing new reserves (although no
one can with much confidence estimate the latter
cost). Oil is still, in this sense, in "surplus",
and it could well remain in surplus for at leat
another decade. This can be expected to give rise
to competitive strains among producers.

On the other hand, the higher prices,
uncertainty of supply, and the fears of the governments of consuming countries for their security have
brought forth large and important investments in
even higher cost sources of oil and/or energy and in
oil conservation. Many of these projects have
already been abandoned and many more will have to be
abandoned if the price of oil continues to fall
nearer to its own long-run supply price - at least
until there is clear evidence that the price is
rising significantly. It is evident that most of
the large and small producing countries and
companies fear a "collapse" of oil prices and are at
present willing to restrain their competitive
instincts to prevent it. This, of course, is no
guarantee that they will be able to continue to do
so successfully for long, but it does indicate that
there will be strong pressures to contain
competition. Just as government regulations of one
kind or another in the 1950s discouraged competition
among oil companies each to enlarge its own share of
the market, so regulations by both importing and
exporting countries may have a similar effect in the
future. Few, if any, of the major agents in the oil
market would welcome falling oil prices.

Mutual restraint by sellers that are few

43

enough in number to appreciate the effects of their own actions on each other and on the market, and where they are also a large enough part of the market to permit them to ignore (or to discipline) a maverick fringe, can be reasonably successful for some time in containing price movements within "acceptable" bounds. If, at the same time, policies of the governments of some importing countries - policies designed perhaps to protect their own higher cost energy-related investments - directly or indirectly support each behaviour on the part of oil exporters, the prospects of success are further strengthened.

If a new system of competitive restraint, of oligopolistic behaviour, of mutual forebearance, is to be established it will probably have to be a tripartite system and, to be effective, must have a "core" to provide leadership. OPEC, and in particular the group of Gulf exporting countries is the only practical grouping to provide such a core. The OPEC countries as a whole are likely to remain the largest suppliers of oil to world markets, with the Gulf countries becoming increasingly important within the OPEC group as the reserves of many of the others decline. They have held together, albeit precariously at times, within an existing well-established organisation useful to them and within which they have been collaborating for twenty years. They have experience under able and respected leaders.

To succeed in such a role, however, OPEC must ensure that its policies are accepted and supported by at least three major groups: the major producers among its own members, the other important exporting countries, and the major oil companies. There are those who contend that from the point of view of pricing and market stability, the oil companies will from now on play a passive role, having to accept and adapt to government policies (providing of course that these policies give them adequate incentives to continue their tasks in exploration, development and production upstream and refining and distribution downstream). There is some truth in this contention but so long as the institutional arrangements governing the sales of British oil through BNOC remain, the large oil companies operating in the British North Sea are in a strong position to make or break an OPEC policy which they are unwilling to accept.

This was evident at the end of March 1983 when BNOC announced a reduction in its prices within the

OPEC and the World Oil Market in the 1980s

OPEC guidelines and at the same time announced that
the companies had accepted the new price level. Had
the companies not done so the whole price structure
would almost certainly have been undermined. For
fiscal, control and so-called security reasons the
British government requires its North Sea oil to be
sold through BNOC. Much of this oil is brought back
from BNOC at BNOC prices by the companies that
originally produced it. They are under no
obligation to buy and they will not do so if they
are able and willing to refuse, buying what they
require on the spot market or elsewhere, or simply
relying on stock drawdown as a means of bringing
pressure. The British Government is committed to a
"market oriented" approach to oil pricing and
intends that BNOC should "follow the market".
       It is, however, evident that "the market" has
changed drastically since 1973. The major oil-
exporting countries have taken over the production
of crude oil which no longer moves in as large
quantities as before through the integrated network
of the major companies from well-head to refiners
and as products into distribution channels.
Instead, government companies sell most of their
crude oil directly and a large and active population
of traders operates in the market. This, together
with a much larger number of companies as buyers and
sellers, had the effect of increasing the amount of
oil and products passing through the spot market.
In consequence, the "market price" became
increasingly less influenced by official prices: the
spot market provides, within wider limits than
before, an alternative source of crude for companies
operating in the North Sea in addition to the crude
they may obtain on long-term contracts with oil-
producing countries, sometimes at less than the
official prices. Thus, in periods when the spot
market is soft companies may have considerable scope
for putting pressure on individual producers, and
notably on BNOC, insisting that the prices set are
out of line with the market. BNOC must sell the oil
it acquires, and although it has some flexibility
(for example it can refuse royalty ol) it has very
little power to resist the companies in such circumstances. Under the present British Government it is
politically necessary that Britain be able in a
reasonably convincing manner to maintain the
appearance of a "market oriented" price policy. But
to "follow the market" does not imply that official
prices must be followed when other prices are under
pressure.

At the same time the British Government seems to have committed itself to the proposition that "stability" in world oil markets is in its interest and in that of the world as a whole. The maintenance of stability in present circumstances and those likely to prevail until at least the end of the decade is likely to require "responsible" restraint on the part of the sellers of oil. But stability is not inconsistent with a downward drift in prices providing that movements are reasonably slow and "orderly" and do not disrupt individual economies, and that decisions are taken quickly in conditions that do not give rise to excessively destabilizing expectations.

Even today (June 1983) the market is essentially dominated by the voluntary restraint of producers (and buyers) inspite of the fact that, in principle, OPEC countries are subject to individual production quotas. The restrain may be effective so long as oil markets are so weak that sellers are acutely conscious of the immediate consequences of competitive price-cutting. But if oil demand begins to recover and if it is accompanied by optimistic expectations of further rapid recovery, the system is likely to come under increased strain as output begins to expand. Individual producers may even begin to think of raising prices to ease some of their financial constraints. At this point collective discipline may be more difficult to maintain, just as it was during the period of OPEC dominance in the 1970s. The strains on the system will be intensified if competition in product markets increases as OPEC countries attempt to sell more products from their new refineries; it will be more difficult to observe restrictions on crude output destined for domestic refineries. And when Iraq and Iran resume exports on a larger scale...?

The task is formidable, requiring both highly skilled diplomacy and sophisticated economic policy. The general level of the "reference" or "marker" price of crude must be compatible with overall supply and demand balance, for the effect of price and of the policies of suppliers on the amount demanded cannot again be dismissed as irrelevant. Differentials around the marker price must reflect the particular competitive conditions in the market for each type of crude oil. Cartel quotas are far too clumsy and blunt to provide effective instruments to resolve the conflict of interests and the potential incompatibility of individual decisions taken on separate issues. A "market oriented" oligopolistic policy adopted by all the major producers might have a chance of reasonable success providing that "gluts" are widely perceived to be a serious potential danger to price stability. It is in this area that skilled diplomacy will be needed to keep the danger at the forefront of producers' minds.

Much of the burden of "leadership" would inevitably fall on Saudi Arabia and the Gulf producers by virtue of their existing and potential capacity. During the periof of OPEC dominance Saudi Arabia's inability effectively to enforce the "moderate" price policy she sought stemmed from her inability to exercise the market leadership consistent with her weight in oil production without comprising her political leadership within OPEC, and especially in the Arab world. The fact that OPEC countries will no longer be able to call the tune on prices will in many ways strengthen Saudi Arabia's position since the smaller producers in OPEC will no longer be as important in the formation of OPEC policy; they will have little choice other than to recognize the greater importance of non-OPEC producers in the market.

I conclude that the major oil-producers of the world, with the tacit agreement of the major international companies, and under the broad leadership of the Gulf oil-producers are likely once again to attempt to observe, in some cases explicitly and in other implicitly, the rules of an oligopolistic game which may, once again, maintain a reasonable stability of oil price for perhaps another decade, but probably not at $29/b for long.

NOTES

1. See M.A. Adelman, The World Petroleum Market (Wash. D.C.: Resources for the Future) Chapter V.
2. See "The Ups and Downs of OPEC Prices, 1960-82", Petroleum Intelligence Weekly, Vol. XXI, No. 15 (April 12, 1982), p.11.

OIL COMPANIES AND THE CHANGING ENERGY MARKET

Sir Peter Baxendell

Ladies and gentlemen,

It is said that experience is the name that everyone gives to their mistakes, and when we talk about the experiences of the past ten years in the energy scene, this rings all too true.

The present confusion and apprehension surrounding oil price prospects must seem, to the general public, to be illogical ... over the past decade commentators have attributed the world's economic and related social problems to the uncontrolled rise in oil prices; today many are showing greater gloom at the prospects of prices falling. Severe problems have been caused by oil supply constraints; now we are told by some commentators that over-supply can be equally unwelcome. Certainly this would be absurd if the so-called oil and energy markets were in fact governed by traditional market forces. The events of the past ten years, however, have brought into the energy scene such a variety of actors, and such a multitude of often-conflicting forces, that it is impossible to assess or understand energy develop- ment in conventional market terms. Single-line forecasting and planning is equally misguided, which worries those used to making decisions on this basis.

Professor Robinson has assessed the changes in the energy 'market' in broad terms; my brief is to present to you the view of a major international oil company which has, of course, been all too closely concerned in recognising and coping with those changes and which has to continue reacting flexibly and operating confidently in the face of current and future uncertainties.

I think any analysis has to start with a look back at the events which have brought the oil industry to its present position - a position which is very markedly different from that of the early 1970s. A few basic statistics show the extent of that change. Excluding the communist areas and North America, the share of crude oil production in the hands of national companies in producing countries increased from 10 to 70 per cent between 1970 and 1979. In the same period, the major oil companies' share dropped from 72 to 17 per cent. At present, anything from 30 to 40 per cent of all crude oil supplies handled by the majors is estimated to come from short-term sources rather than from the companies' own output or longer-term contracts. Ten years ago, the figure would have been 5 to 10 per cent at most. Product sales by state marketing companies rose from 10 per cent in 1971 to just over 20 per cent in 1981, while the major companies' products sales fell from 54 per cent to just over 40 per cent. Thus the position of the international oil companies in the market place has altered radically. Ten years ago, they were feeding equity oil through an integrated system designed to maximise economic supply to growing markets. Today, the same companies are supplementing their own production by buying in crude oil and products, increasingly on a 'short-term' basis, for trading in markets constrained by economic malaise.

1. THE PATH OF CHANGE

Tracing the path of this profound alteration in roles, we may have to go back even more than the 10 years under discussion at this conference.

It began slowly during the 1960s, as many smaller independent oil companies, who were obtaining oil from new producing areas, entered the oil markets and began gradually eroding the dominant role of the major oil international companies. From the early 1970s, the major oil companies found that the volumes of oil available to them under equity arrangements dropped as traditional concession arrangements were phased out.

This process accelerated rapidly as the oil crisis of 1973/74 brought fourfold price increases and the first major taste of politically imposed supply constraints. Nevertheless, at that time the major international oil companies still had sufficient access to crude - including the 'back back' of nationalised equity oil - to be able to cushion the effect of embargoes and shortages by feeding the available supplies through their

integrated systems. They sorted out their own
supply problems, and distributed to customers as
equitably as possible, with a minimum of overall
world disruption.
     But once the shock of the oil price and supply
disruptions abated, and the period of uncertainty
and rapid change was felt to have passed, then as
far as the consuming public was concerned, things
went back to normal. For the oil companies, however, that was not the case. The 1973/74 crisis
began a major shift in the influence over supply and
price towards the governments who had taken over
control of their production. Their policies were
geared, not to straightforward commercial criteria,
but towards a complex range of national and international political and economic objectives. The
obverse of this was that consuming governments also
felt the need to become involved with the oil
market; price controls were extended or introduced,
and often applied in ways which bore no relation to
market mechanisms.
     Later in the decade came the 1979/80 oil
crisis with all its related political tensions and
oil supply disruptions. Consumer complacency was
shattered. Prices leapfrogged; traditional quality
differentials gave way to quantity competition.
Spot prices soared as buyers fought to insure themselves against real or perceived supply shortages,
and those transactions pushed the whole price
structure upwards. The effect on world economies of
this sequence of events is well known to all of us
here - we are still living with the reverberations
of those oil price shocks and the insecurity that
came in their wake.
     This insecurity led many consuming countries
to attempt to secure their supplies by bypassing the
traditional integrated channels and entering into
government to government deals. Politics became
increasingly influential in the oil markets. Term
contracts for oil - previously the cornerstone of
the oil majors' international trading - were
unilaterally changed or cancelled. Premia were
imposed selectively on individual transactions.
     With the benefits of hindsight, it is evident
that the oil crisis of 1973/74 affected consuming
countries' behaviour to differing degress (and some
hardly at all). The influence of this first crisis
on energy demand resulted largely from restricted
income, as economies slowed down, and there was some
degree of psychological effect, mainly on those
governments who saw their supplies threatened and

their balance of payments subject to uncontrollable oil price fluctuations. In the short term, however, the response was inadequate to mitigate the effect of the Iranian Revolution which precipitated the second crisis.

But the 1979/80 crisis boosted the psychological impact and the customer's perception was one of unremitting increase in oil price, which accelerated the trends towards conservation and substitution by other fuels. Thus there was rapid response on both the energy demand and supply sides.

The past few years have seen still more changes. Oil demand has fallen as a result of recession, conservation and substitution - though the degree of effect of each is still uncertain. Oil stocks have been reduced, but substantial excess crude oil production capacity still overhangs the market. Product prices have been under constant pressure, particularly in Europe. Indeed, the only certainty now seems to be that nothing in the oil world is certain. And fear of the unknown is exacerbating existing worries about the future to the extent that any change is seen - by one commentator or another - as the launching pad for even greater disruption.

Inevitably, during the past decade the energy market has become highly 'politicised'. It is of international diplomatic significance and of national economic and social significance. Energy costs and security have, and will continue to have, a major impact on national economic policies, for good or ill. They impinge on almost every aspect of national life, from inflation and balance of payments considerations through industrial costs and structure down to the lifestyles of individuals. Because of the 1973/74 oil price and supply shocks, the IEA was formed to provide an international forum for energy deliberations and, where necessary, or possible, to act on energy problems. Nationally, governments of energy importing countries (and some with domestic surpluses) have become increasingly interested in developing indigenous supplies of energy. They believe that local energy, even in comparatively small quantities and in subsidised developments, will improve supply security, employment and the balance of payments. The current uncertainty on oil prices is causing concern in many of these areas.

2. OIL INDUSTRY RESPONSES

I mentioned earlier that the oil industry has needed

51

to remain flexible and responsive to the changes in
energy markets. Obviously, I can speak only for
Shell companies, but I would like to say a word or
two about changing perceptions over the past decade,
and the measures we have taken in evolving a pattern
of activities to cope with the new energy scene.

Primarily, the industry as a whole has felt the
effects of the increased number of actors in oil
markets, particularly in oil trading activities.
Following the oil supply and price fluctuations of
the '70s came an independent breed of commercial
traders working at the margin. They offered crude
oil and products at whatever price oil-short
consumers would pay, often driving prices higher
than ever and encouraging short-term agreements.
This, together with their decreased entitlements to
equity oil, has brought a changed perspective for
the major international oil companies, and for the
past few years there has been a tendency towards
redefining the upstream and downstream areas of
their activities. In the days when the oil
companies were responsible for the full chain of oil
and gas production from exploration through to the
market place, via transportation, refining and
marketing, then it was possible - indeed part of the
efficient running of the business - for losses in
some areas to be balanced by profits made in
others. Now there is a need for each part of the
business to be viable in its own right, and each
should indeed be treated as a separate business
area, with the links between forged on the basis of
economic and competitive practice.

As you may know, the Royal Dutch/Shell Group of
companies runs on unusually decentralised lines.
This loose-linked organisation has an in-built
flexibility of response which has proved its worth
in times of rapid change and constant uncertainty.
Every Shell operating company is responsible for its
own business, and is expected to prove viable in its
local situation. During times of oil supply
difficulties, for example, the ability of operating
companies to make their own decisions on supply of
crude oil and products without any overriding
central direction has proved advantageous; here, as
in all other areas, the international network of
Shell companies provides a back up of experience and
expertise.

This does not mean, of course, that the Royal
Dutch/Shell Group of companies is going to 'break
up'. Nor do we believe that our business lies in
marginal short-term profit. For Shell companies,

the 'new perspective' has reinforced the belief that
flexibility is a necessary - indeed, a vital -
asset. It has encouraged analysis of operational
effectiveness, and of problem areas. Which, of
course, has meant considerable thought being devoted
to the operations of our downstream businesses,
particularly in Europe. The current situation of
considerable over-capacity both in oil and chemicals
has affected everyone in those industries. Shell
companies have recognised the need to 'slim down'
their activity base; in fact we, like the rest of
the industry, have to recognise that we face a low
growth - or, in some areas, a declining - product
market and work on that basis. Fortunately, the
strategies which have been adopted and implemented
over a number of years - concentrating on technical
expertise and a market-oriented approach - have put
us on the right path for responding to today's
market conditions.

I have already mentioned the need for
flexibility. I cannot say too often that in all
areas of decision making and of operation this has
been our key word for a number of years, and I
expect it to be so in the future. When uncertainty
prevails, and you are involved in a large scale
continuing business involving areas of very high
investment, you must be prepared to assess
possibilities and probabilities, and react to
current events while still taking a responsible
long-term approach. Despite the changes in our
environment, there are still plenty of projects that
require a multinational's scope and strength; there
are also opportunities for us to use our skills
profitably in smaller ways - which also spreads risk
in uncertain times.

Having said all that, I want to make it very
clear that we in Shell accept that we are in a
long-term business making long-term decisions.
Consideration of short-term profitability should
never be the only pointer to future action for a
responsible enterprise capable of mature judgement
of long-term interest.

As I see it, one of the fundamental problems
in today's energy scene is to reconcile the
objectives and priorities of authorities, both
national and international, with the contribution to
a secure and stable energy market which can, to a
significant degree, only be made by commercial
institutions.

'Securing energy supplies' does not just mean
obtaining enough oil, or enough gas ... energy

available in the necessary quantities, at the right
time and at the right place, for continued growth.
It means making energy available at prices which do
not interrupt or prevent economic expansion, and it
also means securing supplies of energy not just for
now, but for an extended period, so that governments
and private industry can plan with confidence. And
the key to this is some degree of predictability in
the oil markets; oil supply and oil pricing should
not be shrouded in fear and uncertainty, with the
energy world periodically jolted from complacency to
crisis and back again.

3. DEVELOPMENTS ON OIL PRICE AND PRODUCTION
Of course, the overriding issue in the energy world
at present is the debate over oil price and
production control. There have been as many
opinions passed on the pros and cons of a fall in
oil prices as there have been commentators holding
forth on the subject. I am loth to add to their
number, but I do have a few points to make in this
context. Firstly, as far as the international oil
industry is concerned, the events of the 1970s have
largely inured us to rapid change and uncertainty.
The major companies have had to come to terms with
successive problems and crises which sometimes
affected us crucially but over which we had very
little influence. We have learned to adapt and for
the most part have done so successfully. We shall
do so again, with more or less difficulty according
to price developments.
        As a matter of common sense, there is clearly
a case for suggesting that lower oil prices will
bring certain advantages. I must say, however, that
I have found no unanimity amongst economists
(perhaps it would be more surprising if there were)
about whether this is really so, or about what
economic advantages would in fact follow a price
drop greater than the $5 we have seen so far.
        Perhaps all we can safely say is that if oil
prices are to fall still further, it is clearly
preferable that this should happen in an orderly,
gradual way. A rapid and disjointed collapse could,
as you will all be aware, cause major international
financial problems.
        Even an orderly decline in prices could,
however, pose potential problems, and these centre
mainly on the longer-term development and use of oil
and energy. If oil prices fall significantly,
public perception could well revert to its 1960s
view of oil as an ever-available, relatively cheap,

all-purpose fuel. Complacency would return and seriously impair efforts towards developing future energy sources and towards conservation and substitution - negating the few benefits which have come out of the oil crises and pushing constructive energy concerns out of the public consciousness.

The development of alternative energies is already in some doubt in certain areas. As we all know, nuclear programmes are under pressure or at a standstill in many consuming countries. But other alternatives, and indeed some high-cost conventional energy sources, also face problems - cost inflation; reluctance to enter into long-term commitment of vast capital sums when the eventual return is uncertain; and the increasing burden of taxation imposed by some governments. Projects are already being shelved or abandoned: solar energy, coal gasification and synthetic fuels; undeveloped oil and gas fields where geology or location pose particular problems or where reserves are relatively small - all these have been affected. New fields in relatively inaccessible areas, such as Alaska and northern waters offshore Norway, may also be subjected to critical scrutiny. Cash flow problems induced by weak oil markets in recent years have meant that some energy investors have already slashed capital spending programmes; should oil prices fall too far, the heavy investment needed to develop other future supplies could appear uneconomic.

I believe that if prices do decline, governments must be prepared to renegotiate fiscal conditions both on high-cost conventional oil and gas projects and on the less conventional alternatives which will provide supplies for the longer-term future. If this does not happen, if investment is not made in good time, and if the impetus to conservation is also allowed to lessen, then we in the industrialised world will be back in the old trap, dependent on supplies from those who may not be willing to make them available in response to consumer demand.

Let us be quite clear as far as oil is concerned. OPEC production has now fallen below non-OPEC production. Nevertheless, the capacity to produce, both in the short- and particularly the long-term, lies overwhelmingly with OPEC. Non-OPEC production has risen more than two-fold since the first oil crisis, but I believe it will soon level out, and thereafter decline. Production from the UK North Sea, for example, will peak within the next

55

few years. The non-communist world's reserves lie predominantly in OPEC hands, and in the absence of alternatives, the consuming world will be dependent on the policies of the major exporters. If consumers abdicate responsibility for the energy future again, we shall almost inevitably face energy constraint again - and if that happens the price roller coaster could shoot back up before we know it. I hope governments will recognise that current conditions are a disincentive to certain necessary developments and will be willing to introduce fiscal arrangements which stimulate investors, rather than deter them. In this context, the changes in North Sea taxation introduced by the UK Chancellor of the Exchequer in the recent Budget are an encouraging sign.

Despite the recent OPEC agreement around a lower -$29- marker price, we still don't know what will happen to oil prices. Nor can we confidently predict the economic effects of the price changes which have already occurred.

Speculation will, of course, continue but I will not. I will simply conclude by saying that new opportunities are arising as the environment changes, and that I believe the major international oil companies such as the Royal Dutch/Shell Group accept their changed role, and are prepared to continue their commitment to developing energy supplies not only for present needs but also for the longer-term future. Speaking only for Shell companies, I think we are now benefiting from the flexible approaches we have developed over the past, troubled decade. Fundamentally, we aim at a capacity for rapid response to short-term change in concert with stable longer-term investment programmes for the discovery and development of new energy supplies.

Whatever the environment, one thing is certain: uncertainty has become a fact of life in the oil world and all of us have to make the best of it. It is a challenge, but I think it is a challenge we can face up to more confidently in the knowledge that we have survived the traumas of the 1970s and learned valuable lessons from them.

OIL SUPPLY AND DEMAND

M A Adelman

I will take a specific and some would say narrow
view of my assigned topic. Demand is what consumers
will take, given their incomes and relative prices.

Under competitive conditions, suppliers too
treat price as an external fact, to which they
adapt. They react to higher prices with more
investment and capacity. But a monopoly makes the
price, it does not react to it. Price does not
determine investment, but the other way round.

In theory, we would draw up a relation between
price and supply under monopoly, and some models
have been built which purport to do this. They
have, in my judgement, been unsuccessful because
they have neither been able to estimate the basic
cost relation, nor to capture the monopoly process.

I intend to fall back on some simpler notions,
to see what can be said about competitive oil
supply. In effect, this means replaying the history
of the last ten years, and asking what would have
happened if demand and supply - the cumulation of
cost - had been left to play out their game
unassisted by OPEC or any other protagonists. We
may not find the answer to that interesting
question, but we may understand something of what
has been happening.

ENERGY DEMAND, PARTICULARLY OIL
In June 1979, the OECD ministers assembled in Tokyo
thanked Sheik Yamani for his promise that Saudi
Arabia would "never" (sic) permit the price of oil
to go as high as $20, and resolved bravely to cut
oil consumption. Since then, there has been endless
debate on the wrong question: how much of the
reduction (from what level?) has been due to
recession, how much to conservation? Since
conservation is a Good Thing, and recession a Bad

Thing, there is heavy breathing on whether the gains
on the Good side are worth the losses on the Bad.
It is all quite unnecessary.

Consumption normalized by income. We can use
the simple equation or identity: C=G(C/G), i.e. oil
use equals volume of economic activity (GNP)
multiplied by oil intensity, or use per unit of
GNP. Thus we can look right through the
fluctuations, to see oil consumption normalized for
recession and recovery. Immediately upon the first
oil shock of 1973, oil intensity declined in almost
every year, accelerating after 1979.

From 1973 to 1982, in the non-OPEC world,
where consumption reacted to prices paid, oil use
dropped eight percent. (See Table 5). Incomes in
the developed world rose 23 percent, in the non-oil
developing countries by 44 percent; perhaps we can
round this off to 27 percent growth. Then oil
intensity declined by 28 percent in nine years, or
not quite 4 percent per year. In the United States,
oil use is down 12 percent, while national product
is up 21.5 percent, so oil intensity is down nearly
28 percent[1].

Two Price Impulses. This decline in oil
intensity has come in two installments. The first
price explosion was in 1973-74. Thereafter real
(inflation-adjusted) prices were steady to lower
through 1978. Yet oil intensity continued to
decline about one percent per year in the U.S.,
somewhat more in the other OECD countries.

The 1979 oil shock was much greater in
absolute terms, hence more important, than the
first. Our usual idea of elasticity, the change in
relative price, plays us false here, when we are
concerned with very large differences. If we think
in terms of substitution, then the amount of
resources worth using up in order to avoid using a
barrel of oil depends on the price of that barrel;
the greater the absolute change, the greater the
shift.

The consumption response, i.e. the change in
oil-intensity, was accordingly much greater after
the second oil shock, and oil intensity declined at
the rate of about 6 percent per year during 1978-82.

The "Polluted" Data Year 1979. It is a
mistake to reckon any trend from 1979, because the
1979 panic saw a large increase in "downstream"
inventories, i.e., those held by distributors and
ultimate consumers. It is shameful that we have no
statistics at all on them.

In a thesis written at M.I.T., Dr. Ellen

TABLE 6

WORLD (excl. Communist areas) OIL CONSUMPTION, SELECTED YEARS 1973 - 82

million barrels per day

Year	1973	1978	1982
TOTAL	47.9	50.9	45.7
excl. OPEC	46.4	48.6	42.7
Industrial Countries	39.0	39.7	33.0
LDCs	7.4	8.9	9.7
SUPPLIED BY:			
OPEC exports	29.6	28.1	16.3
Communist exports	1.1	1.8	2.0
Other	15.7	18.7	24.4

SOURCES: Consumption, BP Statistical Review (annual)
Communist exports, ibid., subtracting
exports to non-contiguous areas, estimated
from CIA, International Energy Statistical
Review OPEC consumption, CIA
OPEC consumption, Central Intelligence
Agency

Burton has estimated that 1979 U.S. "downstream" inventory capacity was approximately equal to total "upstream" holdings. If inventories are normally somewhat less than half full, and are filled almost brimfull during times of panic, then the invisible stockbuilding of 1979-80 might have exaggerated apparent consumption by as much as 10 percent. I would regard 1979 consumption data as hopelessly "polluted" and simply disregard them, making 1978 the basic for comparisons forward and backward. But the invisible stockbuilding was compensated, partly or wholly, or perhaps more than wholly, by drawdowns in 1981 and after. Hence consumption is probably somewhat understated in more recent years.

Furthermore, consumption tends to rise and fall somewhat more than income generally, because industrial production swings more than income, and industry is a voracious energy consumer. For our purposes, these are only minor qualifications.

The decline in oil-intensity is in part substitution away from all forms of energy, in favour of labour and capital by business firms, and in favour of other consumption goods by households. In part, it is substitution away from oil towards coal or other energy forms.

Slow Capital Stock Change. I would not expect oil intensity to keep decreasing at anything like 6 percent every year. In fact, I do not believe that it really has decreased at that rate, because of recent-year understatement. But consumption will keep decreasing for years. Conservation is nothing but investment, tangible and intangible. Because of large increases in oil and other energy prices, the whole capital stock of society is being slowly changed over, from less to more efficient.

Rome was not built in a day, nor a decade. The capital stock changes only slowly, especially during a period of slow growth and very low investment. The average life of the assets of non-financial corporations in the United States, according to some still unpublished studies of the U.S. Department of Commerce, is about 18 years. The variation is very great of course, and we must as always beware of averages, but the general picture seems clear.

We have not yet fully adjusted to the price increases of 1974, let alone those of 1979-80. Nor does the recent softening in oil prices make much difference. As compared with the $5 decrease: gasoline (petrol) costs about $120 per barrel in Europe and $50 in the United States. At the other

end of the spectrum, the expulsion of oil from
electric power generation, and much industrial use,
seems assured. The small decline will somewhat
moderate the rate of decline of oil intensity, and
make the ultimate target level somewhat higher than
it would otherwise be. It will not reverse the
trend.

Economic Growth. Outlook for the Eighties.
This brings me to the other factor: economic
growth. Without exception, every industrial country
shows a sharp down-deflection after 1973. Nobody
will ever know just how much of it was due to the
oil price increase. But the oil price shocks
spurred inflation and balance of payments deficits,
and deflationary government policies to counteract
them. All this, of course, in addition to the
direct "tax" on oil consumption, and loss of
purchasing power.

The resulting unemployment has slowly set off
a wave of protectionism which alarms me like nothing
I have seen since World War II. The 1930s were the
worst of times, and the destruction of the world
trading system is putting us back there.

In this connection, there is a dark side to
conservation. Capital invested to save energy is
not available to increase output. The General
Motors Corporation is spending about $40 billion in
the five years 1980-84; for the whole decade, the
automobile industry in the United States may spend
$100 billions. Yet the capacity ot produce
automobiles will probably be no greater at the end
of the decade than at the beginning. They will of
course be very different automobiles, far more
energy-efficient. But the investment in them is
worth making only as a lesser evil, a way of
reducing the cost of using oil in them.

The less developed countries (LDCs) were for
a time thought to be immune, or at least resistant,
to the decline in oil intensity. That was a
mistake, derived from the pernicious custom of
reckoning with "needs" instead of with effective
demand. Having exhausted their borrowing capacity,
and with exports down because of worldwide slow
growth and recession, LDC's cannot pay for oil
imports without cutting back so severely on imports
of producer goods and materials that growth rates
have declined, which reduces also their oil
consumption.

How long the current economic malaise will
continue, no-one can say. Perhaps things will turn
around before the end of the decade. But there is

61

great inertia in these economic tides; measures taken to alleviate the banking crisis, or unemployment, or inflation, have seriously depressant effects.

The combined effect of lower growth and falling oil intensity has been an actual decline in oil use, which last year was around the level of 1971. Nothing like this has ever been seen before, not even in the Great Depression, when oil use fell by 5 percent from 1933 to 1935, then reached a new peak the next year.

Doubtless oil use will fluctuate. Taking ups with downs, if growth is no more vigorous on the average in the 1980s than it has been since 1973, I think oil use will continue drifiting down more than up.

More vigorous recovery will accelerate the decline in oil intensity, because there will be more investment in energy saving. It is no accident that Japan, which has the least bad growth record in OECD since 1973, has also cut back energy use the most.

Stagnation or decline in oil use assumes no pronounced drop in the price. To anticipate a litte: I think that either the cartel will be able to hold the current level or even raise it, depending on whether they are able to keep ranks, or the consuming nations will by taxes and tariffs keep prices from declining drastically.

OIL SUPPLY AND PRICE IN GENERAL[2]

By now, the notion of demand as a relation not a point, a negative function of price, has been so . well learned that it did not need to be set forth. Supply does need some exposition.

The supply curve of any product consists of the array of all possible sources, from cheapest to dearest. But of all those possible source, many are called, fewer chosen. The competitive supply price is the production cost just barely equal to the price which clears the market. Competitive supply includes that point and everything cheaper. Anything more expensive, and that includes the great bulk of the possible sources, is uneconomic, not worth using. Hence the relevant question is: under competition, what would be the most expensive source worth operating?

Oil supply has so far defied formal modeling, because it is more complex than demand, and basic data are lacking with which to capture the relations precisely. But approximations can be used to answer some limited questions.

For any product, cost includes, first, current operating cost, i.e. payments just adequate to obtain fresh current inputs of labour, operating supplies, etc. Second, it includes investment or development cost, which is the payment necessary to obtain fresh reproducible capital, i.e., the needed return, adjusted for risk. Current operating cost can be expressed as a capital sum, by discounting the expected outflows down to the present. In fact, the usual practice is to do so, and add it to the prospective investment. Hence we will speak only of development investment cost, having let it swallow up operating cost.

"This Industry is Different". Before going on to a type of cost which is peculiar to mineral industries, let me mention the stubborn syndrome: "Competition in this particular industry would generate a price too low to last. Needed investment would not be made, capacity would fall too low, the product would be scarce, and in the long run prices would be higher than they would be if a little statemanship, cooperation, etc., would be allowed to work". That it is special pleading does not make it any less sincere, nor less fallacious.

Certainly there are industries where such a case can at least be plausible, if not necessarily convinving: where there are great economies of scale, and room for only a few firms. A former chief energy specialist for the U.S. State Department thinks it holds for wheat farming, of all industries. This tells much about American policy making on international oil, not much on wheat farming.

Perhaps I may be permitted one such general-ization. Competitive markets work none too well; life is full of hitches and glitches. But for gaudy foulups, waste of resources, and dangerous political consequences, look to private monopoly or public regulation. The motives are generally pure; the road to hell is paved with their good intentions.

Basic Cost Relations in Minerals. Every mineral body is finite. Hence higher rates of output impose higher costs, either by the crowding of men and machines in solid minerals, or the decliningpressure of the reservoir for fluids. Cost also rises with cumulative production; the more has been removed from a deposit, the more it costs, all else being equal, to remove another unit.

It is rational conduct always to exploit the better before the poorer deposit, or section, or sediment. Hence within any given deposit, over time

63

cost rises until it equals price, at which point
there is no longer any mineral rent, no surplus
above bare cost of capital. When cost exceeds
price, there is no further investment, and after a
period of decline, the deposit reverts to the status
of a geological fact; it is no longer an economic
asset. What holds for each deposit holds also for
the whole population of deposits.

There is not and never will be "exhaustion" of
any mineral; one gives over its production when it
no longer pays.

Owners of mineral deposits cannot but be aware
of these relations. Since a given mineral unit, if
removed today, cannot be removed later, it pays to
minimize the cost of removal, and to refrain from
extraction even when the price exceeds marginal
cost, if a lower rate of output will so far lower
cost that total rents or profits are greater.

Furthermore, since the industry goes ever from
good deposits to bad, and from bad to worse, the
reserve is always deteriorating. Investment and
operating cost must rise over time, and price with
it.

In other words, part of the cost of removing
the unit today is the loss of the higher future
prices and lower present investment outlays one
would achieve by lowering the rate of development
today. This sacrificed dual benefit of postponing
extraction is often called "user cost".

Of course, the future benefits must be
discounted, making them less important. The present
discounted value of all mineral rents from a given
deposit must be the same for all points in time. It
follows that the mineral rent per unit - not the
unit price, as is too often stated - must rise at
the relevant discount rate. Otherwise the owner
would be better off shifting production forward or
back through time, until present discounted values
were the same everywhere.

Thus "user cost" must be added to operating-
development cost before the mineral owner can make a
rational decision about how much to invest and
produce in any time period. Of course every mineral
deposit has its own user cost, and the supply
function is made up of the array of all mineral
deposits, each with its own two-layered cost
schedule.

MINERAL SCARCITY: COMPLETING THE THEORY
The theory of mineral depletion was a great achievement, in showing how a market system allocates

rationally between present and future needs. In
mineral markets, coming events cast their shadows
before. If future operating-development costs will
be higher, so will the supply price. Then it is
rational conduct to develop less now, and let the
lifetime of the deposit be longer. The lower
production today raises the price, rationing demand
away from present to future.

Had the theory been generally known and
accepted among decision makers, they would have
understood why mankind was in no danger of driving
blindly off the cliff of mineral shortages. The
"energy crisis" would never have become a powerful
myth, and the world would have been spared much.

But the theory of mineral depletion was
incomplete. It set forth what would happen if
mineral resources became ever more costly to find
and/or produce. It sounded like a statement of
fact, that minerals and other natural resources were
rising in price over time because of the growing
pressure of demand upon a shrinking and deteriorating set of reserves. But by the late 1950s, enough
research had accumulated to show that the prices of
minerals tended on the whole to fall, persistently
and substantially, though prices of "renewable"
resources, especially timber, often rose for long
periods of time.

The price of firewood rose by a factor of
three from 1550 to 1650. I have for years risked -
and sometimes received - the derision of students in
asserting that one reason people migrated from Old
to New England was to keep warm. But there are
documents, in the language that used to be spoken in
this island: "Here we have plenty of fire to warm us
... All England, nay all Europe, hath not such great
fires as we have in New England ... Here is good
living for them that love good fires".

Not only did minerals prices fail to rise;
there was no evidence of mineral production ever
being held back when prices covered operating-plus-
development cost. As Richard Gordon put it, the
mineral industries acted as if user cost were zero,
and in retrospect were proved right.

Once the problem was raised, the answer was
obvious. The depletion theory was the truth but not
the whole truth: it omitted the substitution of
knowledge for minerals: knowledge of the Earth's
crust, and of better methods of extraction. For if
exhausted mineral bodies were to be replaced by new
deposits costing no more than the old to find and
develop, there was no reason to hold back

production. What we had in effect was a constant
struggle between the stinginess of nature and the
inventiveness of mankind, with mankind - so far -
getting the better of it.

But "general principles do not decide concrete
cases". Although there is no general tendency for
minerals prices of rise, a particular mineral at a
particular time and place might be getting much more
scarce. Furthermore, the two layers of cost we
mentioned might be moving in contrary directions.
New mineral bodies discovered might be ever more
poor, but improvements in extraction cost might
overbear the deteriorating quality. Or existing
mineral bodies might be so large that the increasing
scarcity might be imperceptible or unimportant, the
slope too mild to bother with.

Finally, the most hopeful and as it happened
the most important case: costs declining, and prices
with them. Then the quicker one produced, the
better. This did not produce any wild stampede to
flood the market, because more intensive development
imposed higher costs, hence a braking or cutoff
mechanism on higher production.

In short, user cost may be substantial or
trifling, positive or negative. There is no
substitute for detailed empirical work. Resources
are limited, but the only important question is:
where are the limits, and how long and steep the
slope to reach them?

APPROXIMATING USER COST
As a way of approximating user cost at the Persian
Gulf, fourteen years ago, I assumed zero discoveries
during 1970-85, while output expanded at 11 percent
per year, consuming 170 billion barrels over 15
years. The growing pressure upon the decreasing
reserves would, I calculated, put the development-
operating cost up from about 10 cents in 1968 to 20
cents in 1981. This was a large increase
relatively, and very small absolutely, or even in
relation to the 1970 price. The increased cost was
what I called Maximum Economic Finding Cost. In
view of the considerable errors of estimate, there
seemed no point in discounting.

(You may imagine my astonishment at reading,
in a paper presented at an OPEC seminar, that in
reckoning oil production costs I had neglected the
fact of mineral depletion. See The World Petroleum
Market, pp. 6 and 69-77. I have written my share of
dull sleep-inducing prose, but had not supposed it
was quite that dull).

More recently, Dr. Nadim Pakravan has made
econometric estimates of Middle East user cost, in
an unpublished paper. They are overestimated, in my
opinion, because of the unwarranted inclusion, in
capital expenditures, of outlays both on the
gas-gathering system and on gas-liquids plants.
Even so, his estimates are only a negligible
fraction of price ..

We can make a rough check on those estimates,
using data for Saudi Arabia. Changes in rig time
per well drilled are a good proxy for changes in
real investment per well. This is not to say that
rig hire is the only investment outlay; only that
investment is proportional to rig time. Production
per well, multiplied by number of oil well
completions, is a good proxy for additional capacity
put in place. Rig time per additional barrel of
daily capacity has risen about 50 percent from
1966-68 to 1978-80. We include all rigs, including
those for water injection and gas, on the theory
that they are all necessary for the additional
capacity; this exaggerates, but not by much.
However, we do not capture the costs of the large
gas gathering system, as we should not.

This suggests that real costs increased by
about 50 percent in 12 years, which is less than we
calculated. That is as it should be: there have
been considerable reserve additions (as against out
estimated zero), and cumulative Persian Gulf output
1970-81 was only 81 billion, less than half of what
we assumed. If we allow for higher factor prices,
both because of general inflation and price movements in oil development, by a factor of about 5,
user cost was forecast at about 50 cents, but turned
out to be only 25 cents.

SAUDI AND LIBYAN ESTIMATES OF VALUE OF IN-GROUND OIL
The 1976 agreement of Saudi Arabia with Aramco
specified a payment for additional discoveries,
which was not tied to the cost of making those
discoveries. Thereby it set out the value to the
resource owner of an additional barrel of reserves:
6 cents. At about the same time, Libya made a
similar arrangement, agreeing to pay 15 cents.
Since the reserve base is much lower in Libya, one
would expect the value to be much higher.

The value of an additional barrel is exactly
equal to the cost of losing an existing barrel. But
since the 1976 agreement must have contemplated a
more rapid depletion of reserves than has in fact
occurred, the 1983 user cost, if calculated, would

be less than six cents, all else being equal.

It is a sobering thought that some people have hung the world price structure on user cost and "scarcity rents". They are worth - at the owners' appraisal - about 2 tenths of one percent of the current price for Arab Light. With apologies to Winston Churchill: never have so many made so much of so little.

This very low value of user cost is consistent with what we know of Aramco production plans in 1973. They contemplated expansion from about 7 million barrels daily to 20 million barrels, with no hint they expected a significant rise in costs. At that time, reported reserves were around 100 billion barrels; Aramco measured them at 245 billion[3].

Production of 20 million barrels daily would have been not quite 3 percent depletion of the fields, a highly conservative rate. It is also consistent with the under- or non-utilization of known fields. At the end of 1980, Aramco was producing from only 15 fields out of 49 known; in the absence of development, there is no telling what the contents of those undeveloped fields might be. Nobody doubts that the country is mostly unexplored, and may contain much mroe than the estimated 245 billion. But there is no point in large scale exploration when there is so little chance of using what is already known. Presumably the more intensive development, to 20 mbd or more, would raise development cost, hence user cost. There is not much doubt that those costs would be only a tiny fraction of costs incurred elsewhere in the world.

SUPPLY CURVE AND SUPPLY PRICE: FROM 1970 TO 1983
Let us now try to make a rough estimate of what the supply curve would look like now, by stepping back to 1970, when the price was about $1.25 for Arab Light, or about $3 in 1983 prices[4].

Since there had for many years been a chronic surplus at the Persian Gulf and elsewhere, the true supply price must have been less than $3. In such of the internal documents of the oil companies as have come into the public domain, we can see how the companies dreaded the demands of the governments for higher output rates, since they saw no way of selling more.

This is confirmed by development cost no more than 10 cents, user cost not much over 6 cents in current prices, which we might as well round up to 20 cents in 1970 prices, 50 cents in today's

prices. Had the whole world market been one
competitive unit, Persian Gulf production would have
expanded, at higher costs; U.S. output would have
contracted, lowering costs there. Thus the supply
price would have been in the range $3 to no more
than $5, with very large competitive rents over and
above the cost of capital ...

We now ask how much the whole supply curve has
shifted up or leftward since 1973. It cannot be by
very much. True, billions of barrels have been
consumed out of the old reserve base. But the
greatest depletion has come in the high-cost areas,
where it has paid to find and develop new reserves.
Very little of the old low-cost reserves of the
Persian Gulf have been used up, because the high
prices dampened demand, and will restrain it even
more in the future. Higher prices and restricted
output have aggravated the basic surplus. Nothing
could be more commonplace and expected. The pattern
is at the opposite end of the spectrum from the
competitive price increase in natural gas.

Reserves worth developing at $5 or lower are
hardly less than in 1970, and may be larger. Those
in Saudi Arabia and Kuwait have hardly been
depleted. Those in Iran are a little less; in Iraq,
considerably more. A good deal of the North Sea is
in that category, and all of the new Mexican
reserves. Of course these costs refer only to the
necessary outlays at the margin, including a
competitive rate of return, with substantial rents
fo the lowest-cost.

Coming closer in time: the price in 1978 was
$12.50, in 1982 prices around $15. There is simply
no chance that the equilibrium price could be any
higher. Indeed, at that price large deposits of
heavy oil in Canada and Venezuela would be worth
developing, and they would put a cap on the price in
a competitive market.

WHY THE UNDER-DEVELOPMENT OF LOW-COST RESERVES?

Why should it be the case that since 1973, the
higher cost reserves have been extensively depleted,
but nearly all replaced by intensive drilling. In
the OPEC countries, some smaller producers have
expanded considerably, but the larger ones have
stagnated; Saudi Arabia during 1979-81 was drilling
only half as many wells in 1973, even when the
collapse of Iran was creating the vacuum nature is
said to abhor. For OPEC as a whole, the expansion
of capacity in recent years has been far below 1973.

We remarked earlier that a basic assumption of

mineral economics was that a rational economic
system depleted the best resources first. And this
would imply not only depletion of known deposits,
but the search for new prolific deposits which would
be overwhelmingly concentrated in the neighbourhood
of the old ones, where chances are best.
     But under monopoly water runs uphill. A
single seller or a group acting in concert do not
react to the price: they make it. If a higher price
and lower output suits their interest, they charge
it and produce less to protect the high price. This
is the only way to explain why the producing
countries with the largest lowest cost reserves have
held them back so effectively.

     The Consensus. Yet on the morrow of the first
price explosion, a consensus formed almost
immediately, which has guided policy every since.
Briefly: there was an energy crisis, but not because
of lack of physical supply or even economic supply.
It was rather that the large exporting nations
preferred to keep their oil in the ground. They
wished to produce only enough to meet their revenue
"needs", especially their development plans. That
could be and probably was much below consumers'
"needs".
     There has been much asking of rhetorical
questions: why would oil producers sacrifice their
interests, and produce more than was good for them,
merely to serve the unbridled appetites of consumers
(above all, American consumers) for oil?
     This aside, the policy problem for consumer
nations was seen as: to curb demand and expand
supply enough to say within the limits set by the
producer countries' preference for keeping oil in
the ground. If we could restrain imports from the
OPEC nations to below the line, we were saved. To
attempt to import more would either open a fatal
"gap" or (to the more sophisticated exponents) would
force prices up. There was a deadly danger, because
higher prices would satisfy the revenue needs of the
producers at a lower production level, which would
make them cut back production, which would force
prices even higher ... one shuddered to think of the
outcome.
     The consensus looks a little shaky today, but
it will come back roaring if there is another
shortage. It amounts to a theory of rational
action, but which excludes monopoly. Each
individual producing country has a constraint on its
production, which is tighter than mere economic

self-interest: the desire to keep oil for the future regardless of whether below-ground wealth would be maximized.

This supposed constraint is quite independent of what other seller nations do. Unlike a collusive agreement, where the production restraint of each seller is strictly contingent upon the restraint of every other, and makes no sense if others don't hold back output, this kind of constraint binds each seller at every moment.

Each nation, managing its assets in its own interests, not counting on the cooperation of any other, prefers lower rates of production than profit maximizers or present-value-maximizers would seek. Hence the danger of runaway prices, and the irrelevance of customary economic analysis. Certainly this is what the foreign offices of the consuming countries have believed, and enlightened public opinion generally.

In fact, the consensus is a tangle of confusion and error.

THE CONSENSUS EXAMINED

Non-economic objectives. This confuses ends with means. The government of a mineral-owning nation may prize some objectives more than material wealth. But no such objectives are sacrificed by maximizing the present discounted value of the oil deposits. There is no tradeoff. On the contrary, wealth is a means of promoting the objectives. More is better, whether for consumption, investment, armaments, influence, or whatever.

Even in the "low absorber" nations, the population enjoying benefits learns quickly to take them as a right, and a regime wanting to stay in power ignores them at its peril. But assume the rulers wisely want to shield the ruled (but perhaps not themselves) from the temptations and distractions of higher living standards. They can park assets abroad, and in fact have done so. There is of course a risk of confiscation, which can be reduced by hiding ownership. The net risk must as in every other investment be compared with risks and returns elsewhere, of which more below.

Backward bending supply curves. We pointed out above that under monopoly higher prices led to reduced output. There is another important type of backward bending curve. In the United States, Canada, Australia, and the North Sea countries, the two price exposions set off the sequence: higher prices --→ higher prospective oil company profits

--→ resentment --→ taxes or price controls --→
lower investment --→ lower output. Thus the
sequence was parallel to what happened in the OPEC
countries, where higher prices resulted in lower
investment and output. Canada is perhaps the
clearest example. In the North Sea countries, since
the oil companies were all leaseholders on the
public domain, it was much more clear to the public
that the result of too-high tax rates was lower not
higher revenues. In the United States and
Australia, the lesson was learned more slowly. In
no country was it learned completely.

But the backward bending curve in the
consensus has been extended from the analysis of
labour markets. It has long been observed that with
rising living standards, work hours decline; hence
the higher the price of labour, the less labour is
offered. But the analogy is false. A worker must
always trade off leisure against income. As his
income grows, he values his leisure time more
highly. But there is no such tradeoff here; lower
income brings no additional leisure or other
blessing. The more income from the mineral property
the more leisure or anything else.

Waiting for Higher Prices. Another alleged
basis for preferring to keep oil in the ground is
the expected higher price in the future. This is a
misapplication of the user cost concept discussed
earlier. The price is not taken but made by the
cartel nations. But let us ignore this for the sake
of the argument.

We have already seen that holding back output
makes good sense, but only to the point where the
cost saving today on lower output rates, and the
discounted higher revenues tomorrow, are greater
than the higher sales revenues sacrificed today.
The smaller OPEC producers, with small reserves,
benefit most by restraining output. Their costs
rise more quickly with higher output, and the
postponement of output is less; hence the effect of
discounting is less. Yet they who would gain the
most by voluntarily holding back output, are the
ones who insist on the quickest production - which
suggests the effect is pretty weak. In any case,
for a very large reserve holder like Saudi Arabia,
the barrel non-produced today will stay in the
ground for 50 to 100 years, its present value is
zero, and we have already seen that their actions
show that they realize this.

Alleged Low Rate of Time Preference. A
personal rate of time preference is irrelevant to

the best use of a mineral property, or any income-yielding property for that matter. The only rational objective is to maximize its present value, discounting the property's revenue stream by the risk factors appropriate to that property, which are objective facts: the probability of receiving income, the fluctuations in expected income, and the effect of those fluctuations in moderating or exacerbating the fluctuations in whatever additional assets the owning party has. (Hence Saudi Arabia, with a large portfolio, would probably though not certainly discount at a lower rate than Nigeria, which has little or nothing). These factors do not change because one rational individual would rather have safe low-yielding assets while another would prefer risky high-yielding assets.

If the Saudis prefer to put their foreign balances into safe low-risk assets, that does not make their oil in the ground a safe low-risk asset! Hence it makes no sense to discount their prospective oil revenues by the low rate of interest obtainable on safe assets, and conclude that oil in the ground, not to be produced for years, has a high present value. Discounted at even a moderately risky rate, the oil of the OPEC nations is not worth keeping underground for possible higher prices. Were each nation to make its own independent decisions, it would produce more, to give them command over whatever bundle of assets they preferred. But of course higher production would undermine prices, and there's the rub.

Government Discount Rates and Private Discount Rates. The effect of the discount rate on the optimal depletion rate is not simple. Low discount rates are a reason for postponing output - far off receipts become more valuable - but also for speeding up development investment, hence output, because the cost of capital is lower. The net outcome depends on particular facts, and seems rather insensitive to discount rates in the likely range. But extremely high discount rates are not ambiguous at all; they choke off investment and make production dwindle[5]. Hence even were it true that oil exporting governments have lower discount rates than private companies, it would be irrelevant.

In fact, oil exporter nations deceive themselves if they do not use rather high rates of discount, if we keep assuming rational action, because they are unstable societies and unruly to each other. The ghosts of Ben Bella, King Idris of Libya, the Biafra rebels in Nigeria, King Faisal of

73

Iraq, his executioner Kassim, the Shah of Iran,
Soekarno of Indonesia, will shortly be joined by
others. They should all serve to remind us how
uncertain is life in the exporting nations.

Producing Only "Enough for Our Needs". If
there were anything to this idea of limited "needs",
expenditures should speedily approach a limit in the
oil exporter nations, and increasing percentages
pile up as foreign assets. The near-panic in 1974
over floods of "camel money" swamping the world
banking system has been too quickly forgotten. The
facts are quite to the contrary: there was such a
hunger for foreign goods and services that surging
imports took the OPEC current-account surplus from
$84 billion to zero between 1974 and 1978; from $120
billion in 1980 to negative in 1982.

Common sense should tell us that "needs" are
what people think they can get. In 1973 the Saudis
were taken very seriously in asserting that all they
"needed" to produce was about 3 million barrels
daily, which at the then-current price amounted to
$2 billion annually. Any production over that was a
"sacrifice" on our behalf. This year, their "needs"
are around $75 billion; at least that is what they
will spend, thereby going into budget and
balance-of-payments deficit.

Indeed, this is one respect in which
governments cannot act as rationally as
individuals. It has proved extremely difficult for
them to live with lower revenues, whether because of
lower prices, lower sales, or both. Reduction of
expenditure means to cut off income from domestic
groups which have grown accustomed to their incomes
and will intrigue or fight to keep them.

WORLD DEMAND AND SUPPLY: A SUMMING UP

What would have happened if there had been no cartel
launched in 1970, and competitive demand and supply
left to work themselves out?

Had price remained at 1970 levels, consumption
would probably not have continued to rise at 7
percent per year. North American consumption, where
the oil-consuming economy was more mature, was
rising at about the same rate as GNP, or a little
more, 4 to 5 percent per year. The switch from
other fuels was accomplished, and the population was
on wheels. At seven percent, use doubles every 10
years, at 4.5, the increase is only 55 percent. In
short, we must make some calculation of consumption
increase at stable prices, instead of simply extra-
polating. But it would have been a massive

increase.

There might have been such growing pressure
upon reserves that actual operating-development cost
and anticipated operating-development cost, which is
the basis for user cost, would have risen
considerably, and price with it.

But had production continued to grow, reserves
would have expanded too, we cannot say by how much.
"Proved reserves" would have kept increasing, even
with zero discoveries, because known but undelimited
reservoirs would have been "booked"; recall that
Saudi Arab proved reserves were 100 but would have
been reckoned at 245 using customary methods of
reckoning. It seems reasonable to suppose that in
every basin the average newly discovered field would
have decreased in size, but by 1985, as my
calculations showed, the effect of Persian Gulf
costs would have been very small even if there had
been zero discoveries.

About all we can safely say is that although
1970 user cost was low because reserves were so
great, hence with little effect on price, 1985 user
cost might have been very substantial. Conversely,
a good finding record might have held costs to where
they were in 1970, if the pact of development had
continued or accelerated after 1973. We will never
know how much might have been added to reserves
either in new or old fields, whether the rise in
development cost and user cost would have been
large, small, or nil. Had Saudi production expanded
toward 20 million barrels daily in 1985, say, costs
might be very much higher today - say one or two
dollars rather than 50 cents - and user costs 60
cents instead of 6. Elsewhere, there would have
been corresponding increases. But then too there
might have been a strong discovery effort mounted in
other parts of the peninsula, and elsewhere in the
great producing areas.

The price explosion, by slowing economic
growth, slowed almost to zero the constant growth in
consumption independent of price (i.e., the
rightward shift of the demand curve). The
competitive supply curve is not much changed from
1970. Little of the very low-cost reserves have
been consumed. In Iran, some of those reserves have
been immobilized, for as long as it takes to restore
some secular reason. There have been some important
additions in Mexico. The total amount of ultra-
cheap oil, costing say less than $3 at current
prices, is larger today than it was ten years ago.
It will stay that way, because the world's

consumption is being supplied increasingly from the high-cost reserves.

PERMANENT SURPLUS, PERIODIC SHORTAGE

The chronic glut will remain indefinitely, and possibly even increase. It is symptomatic of the tension between competitive supply, on the one hand, and the actual fixing of price, on the other.

The permanent glut has not prevented temporary shortages, because the control of production by the cartel is so clumsy, and this is no reproach to them. One cannot fine tune with coarse instruments. If at any moment, the aim is to keep the price stable, then planned production should equal expected consumption. But the managers have only a poor idea of what consumption will be, and they are bound to overshoot or undershoot. But an overshoot will put downward pressure on prices, which they will try to resist by cutting back production. Taking in the slack rapidly, as every sailor knows, means an acute danger of getting a dangerous strain. In 1973 and again in 1979, there were deliberate production cuts to put up the price – the Iranian revolution was at best the occasion or pretext. The reduction was not great, and the deficit could easily have been met out of current stocks. But the fear of dearth, and the hope of speculative gain, produced a surge of demand for hoarding, and put up spot prices; whereupon the producing nations put up prices to "follow the market", which they had themselves manipulated to produce the panic.

These crises cannot be expected to disappear. The instabilities have nothing to do with competitive supply and demand, and everything to do with a very clumsy cartel.

I will not try to guess at the fate of the cartel of the OPEC nations (the organization itself is of no importance). There is certainly a good chance, though I for one would put it at considerably less than 0.5, that it will fall apart. But the world can scarcely return to 1970, even if demand and supply functions allowed it, and they might. Whatever the reasons for the cartel's success, the consuming nations cannot in common prudence allow again the kind of dependence that existed then. One way or another, they will keep their interior price levels high enough to protect interior energy sources, of which the most important is negative: investment for conservation.

Hence the pressure upon world prices will be

all the greater. It would be best for the consuming
nations if there were a grand divorce of price
levels; the lower the price in the world market, the
better, because it would induce more investment,
capacity, and output of low-cost oil.

There is no paradox in lower prices inducing
more output, nor in higher prices inducing less
output. It is a normal monopoly result. But it is
most likely that world prices will be kept above the
competitive level. Large deposits of low-cost oil
will lie fallow for many years, perhaps forever.

NOTES AND REFERENCES

1.   For incomes, see Economic Report of the
President 1983, p. 285; U.S. oil consumption,
Monthly Energy Review January 1983 p. 30.
2.   The following is a very brief
condensation of the received theory of mineral
depletion and user cost. The classic papers are by
Lewis Gray (1913) and Harold Hotelling (1931).
Equally important, in my opinion, were the challenge
made by Barnett and Morse (1963) and the response by
Gordon (1967) and Herfindahl (1967). For an
excellent summary, see Herfindahl & Kneese, Economic
Theory of Natural Resources (1974). The later flood
of contributions is too voluminous for reference.
Much of it is addressed to problems of inter-
generational equity and other issues of welfare
economics, which do not concern us.
3.   Multinational Corporations and United
States Foreign Policy. Hearings before the
Subcommittee on Multinational Corporations of the
Committee on Foreign Relations. United States
Senate. 93 Congress, 2 Session. (Washington,
1974). See pages 360, 366, 442, 452-3, 519, 537,
539. Note, however, that the Iranian Consortium was
confident of its ability to go from 6 to 10 million
barrels at only a mild increase in cost. This
proved wrong; 7 million was the limit. See Maureen
S. Crandall, "The Economics of Iranian Oil", M.I.T.
Energy Laboratory Working Paper 75-2.
4.   As previously, we use the implicit
deflator for the U.S. GNP, which was 210.73 in the
last quarter of 1982, and 91.45 in 1973.
Alternatively, we may consider that drilling costs
rose by a factor of approximately five during that
period. But much of the increase was the result of
a frantic drilling boom which subsided in 1982-3.
Much of it will be reversed, or has been already;
but we cannot estimate some kind of normal price

level for drilling services and components.
        5.   M.A. Adelman, "OPEC as a Cartel", in
Griffin and Teece, eds., <u>OPEC Behaviour and World
Oil Prices</u> (1982).

THE FUTURE OF NUCLEAR ENERGY

Professor Wolf Häfele

Mr Chairman, Ladies and Gentlemen,

I speak with some hesitation because I am a
physicist and an engineer but not really an
economist, but even physicists and engineers have a
view of the economic future of nuclear power and I
will try to express myself in the language that is
perhaps understandable to you. When speaking on the
future of nuclear energy, I think it would be
worthwhile to consider the present state of nuclear
power as there are a number of features in it that
are unknown to many people. I will try to elaborate
on this.

A first look (Figure 3) at the total capacity
of nuclear power stations operating in 1982 throughout the world shows that the total sum amounts to
nearly 168.4 GWe, and this is not a small capacity
(nearly twice the total power capacity of the
Federal Republic of Germany). If we further regard
the increase of nuclear power capacity since 1970,
one comes to the conclusion that this increase has
been more or less steady and linear. This is a
surprising fact and in contrast to the general
impression because there were no interruptions or
breaks in it for the world as a whole.

If we then consider the next ten to fifteen
years and look at the nuclear power stations under
construction and already ordered, (Figure 4), we
will find that this sum is twice that of the actual
operating stations I mentioned above. Taking a more
detailed look at the power plants in operation (left
column we see that 60 nuclear power stations are
operating in the USA, 59 in Western Europe - that
means more or less the same number - 22 in Eastern
Europe that means the Comecon area, and 28 in the
rest of the world, mainly in Japan. The second

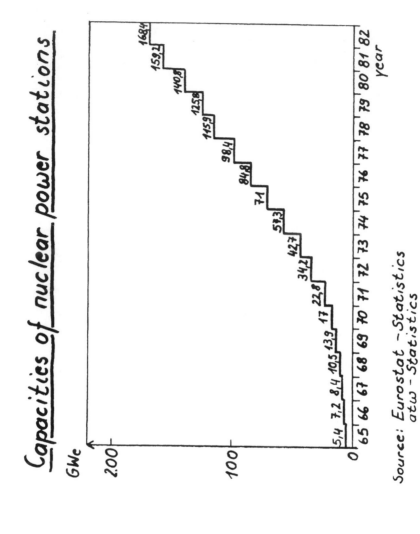

Capacities of nuclear power stations

Source: Eurostat - Statistics
atw - Statistics

Figure 3

Current capacities
of nuclear power stations, worldwide
(as of August 1982)

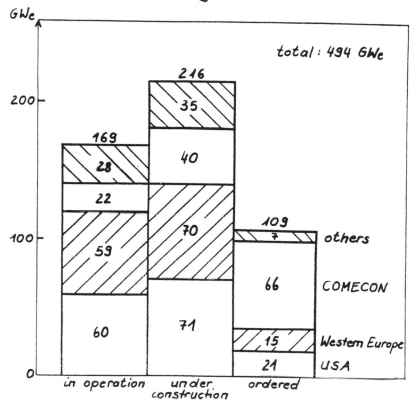

Figure 4

column in Figure 4 shows that in the USA more
nuclear power stations are under construction than
operation, namely 71, and nearly the same number
also in Western Europe, another 70 power stations.
In the Soviet block there is a relatively large
number under construction compared to the number
already in operation. Most of these power stations
under construction in the Comecon area are of the
Woronesh 440 type, the first reactor which will be
built in mass production and is no longer burdened
by technical modifications. Finally, a further 35
nuclear power plants are under construction in the
rest of the world, again mostly in Japan.

If we look at the nuclear power plants ordered
today, the USA with their 21 plants occupy a very
low position in the world league and I will come
back to this a bit later. Also Western Europe has
ordered only 15 new plants, whereas the Comecon area
dominates with its 66 newly ordered plants, (the
right column of Figure 4). The important feature of
this figure is therefore, if we look at the future
development of nuclear energy more generally, that
its centre of gravity for nuclear power moves away
from the USA. It is shifting to Western Europe and
even more explicitly to Eastern Europe, because all
the nuclear power of the Comecon area is
concentrated west of the Urals.

If we sum up all these numbers then we can
expect an installed nuclear capacity of almost 500
GW worldwide in the next 15 years. This corresponds
to a thermal capacity of about 1.0 TW or almost 15%
of present world primary consumption, and is
equivalent to 14 million barrels of oil per day
(bbl/d). This implies that when all these nuclear
power stations are in operation - and no doubt they
will be, except possibly for the 21 of the USA but
these are minor compared to the total - the
contribution of nuclear power energy supply will
essentially parallel the present oil production of
OPEC. I do not think that this is fully recognized
worldwide. And if we suppose that overall energy
consumption will grow in the meantime, the nuclear
contribution will be in the order of approximately
10 to 15% of the total energy produced worldwide at
that time. I conclude from such numbers, that
nuclear power is present and will not easily vanish.

As far as my own country is concerned, there
were three new orders last year. There have been
new orders in France, and in Japan. Indeed, there
is strong support for nuclear power in these three
countries and the increase of worldwide nuclear

capacity is mainly attributable to these. It is
also worthwhile to have a look at Sweden, because
Sweden is said to be anti-nuclear. Well, it has
twelve nuclear power stations generating 50% of all
its electricity. Since almost all the remainder is
hydrogenerated, there is almost no electricity
generation on a fossil basis in Sweden. Similar
numbers hold for Switzerland. These are facts that
are not well-known worldwide. That nuclear energy
has this relatively strong position is a consequence
of the very fortunate price picture which will also
remain in the future.

Let us thus take a closer look for instance at
the most recent results of the Energiewirtschaftliches Institut in Cologne, (Figure 5), where
Deutsche Pfennige/kWh have been transferred into US
mills/kWh because it is a unit that is probably
better known. Here, under different price
assumptions, comparisons of electricity production
costs for both a nuclear and a hard coal-fired
station, both to be ordered in 1982 and to be put
into operation in 1989, are plotted as a function of
the annual load factor. In these stormy days one
can assume an average inflation rate of 4.5% per
year during construction time as well as a price
increase for domestic coal of 5.5% per year and for
imported coal of 7.5% per year on average (see
Assumption A). It follows that even if imported
coal were to remain relatively cheap, nuclear
electricity generation would have cost advantages
over hard coal at load factors above 3500 h/a. In
the case of unfavourable coal price developments,
according to Assumption B, the cost advantages of
nuclear energy rise to more than 40 mills/kWh at a
base load of 6,500 h/a. This is a cost comparison
which the potential power station purchaser should
keep in mind today.

In my own country only the brown coal-fired
stations have electricity production costs
comparable to those of nuclear power stations
because brown coal has been extracted here from open
pit mining mainly in the area between Cologne and
Aachen.

If we now come back to the actual situation in
the USA we should ask what has happened there. I
already mentioned that in the United States the
orders are very low and the problems are very high
and therefore it is perhaps revealing to study and
consider the case of the USA in greater detail so
that we can draw certain conclusions from it.

If we look at the annual orders and not only

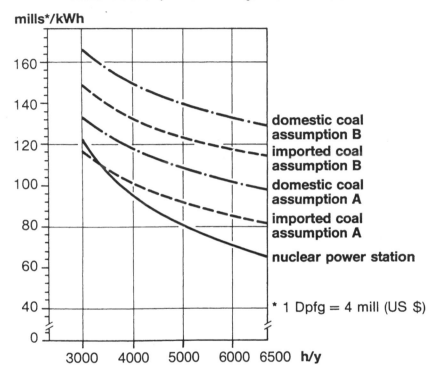

Electrical power generation costs (busbar) for coal fired (hard coal) and nuclear power stations as a function of the annual load factor, start of operation 1989

mills*/kWh

domestic coal
assumption B

imported coal
assumption B

domestic coal
assumption A

imported coal
assumption A

nuclear power station

* 1 Dpfg = 4 mill (US $)

3000   4000   5000   6000 6500   h/y

Assumption A: increase of domestic coal 5.5% / a
increase of imported coal 7.5% / a

Assumption B: increase of domestic coal 6.5% / a
increase of imported coal 9.0% / a

Average inflation 4.5% / a

Source: Energiewirtschaftliches Institut an der Universität Köln

Figure 5

the orders but also the cancellations of nuclear
power stations in the USA (Figure 6) we will see
that since 1979 there have been no new orders but
only cancellations. Last year another 16 power
stations were cancelled, and this indicates that
they are on the way down and out. This is a very
drastic picture which is very painful to me and it
means that in the USA we can expect a large although
static volume of reactors, around 130, and that the
related industry will concentrate on operation and
maintenance, but not on further development. Again
we ask, what has happened in the seventies which
causes the American electricity suppliers to withdraw from the nuclear power market?

     The Electric Power Research Institute (EPRI)
in Palo Alto has investigated the causes of the cost
explosion for nuclear power stations and the
resulting unwillingness to invest in the USA. Even
if the results cannot be transferred in detail to
other countries they nevertheless do provide an
initial survey of the influential factors whose
importance can be examined for each individual
country. Figure 7 shows the various factors of
influence on cost increases for coal-fired and
nuclear power stations from 1969 to 1978 and,
adjacent to this the respective cost structure of a
power station ordered today and to be put into
operation in 1990.

     The 1978 costs are composed of the original
1969 costs, plus a percentage (250% in the case of
nuclear power) which is made necessary by
modifications to the regulations and requirements in
the meantime, a further percentage due to inflation
(almost 150% in the case of nuclear power), and
finally an additional sum (another 100%) to cover
unforeseeable cost factors only becoming effective
during the construction period (such as strikes,
interventions, backfitting costs or delayed
commissioning). A direct comparison between coal
and nuclear energy shows that in the case of
coal-fired power stations particularly the
modifications of the regulations (e.g. tightened
environmental requirements) caused the decisive cost
increases, whereas in the case of nuclear power
stations all factors made an almost equal
contribution to the rise in costs. If in the case
of costs for a facility ordered today to go into
operation in 1990 one presumes that the standards
and construction regulations were not to change
significantly, then the growth rates of the specific
capital costs for coal-fired and nuclear power

Annual orders and cancellations of nuclear power stations in the U.S.

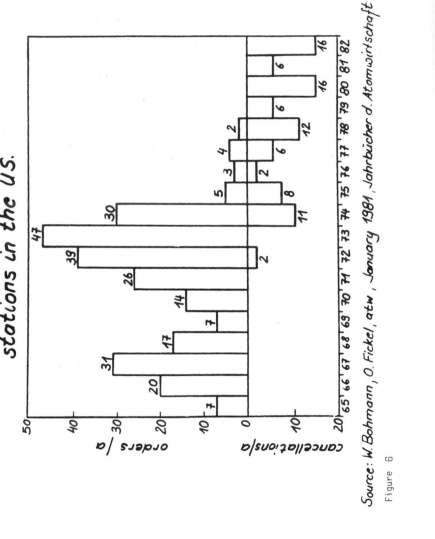

Source: W. Bohmann, O. Fickel, atw, January 1981, Jahrbücher d. Atomwirtschaft

Figure 6

Impacts of various parameters on the cost increases of coal and nuclear power stations

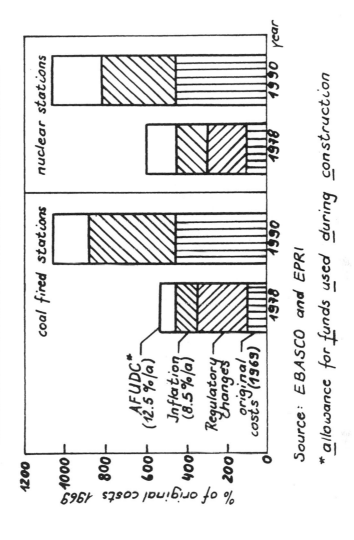

Source: EBASCO and EPRI

* <u>a</u>llowance for <u>f</u>unds <u>u</u>sed <u>d</u>uring <u>c</u>onstruction

Figure 7

87

stations are approximately equal, that is to say
they exceed the original 1969 costs by about a
factor of 11. The interesting point is that the
high cost increases are not a specific feature of
nuclear engineering. Their origin is not to be
found in the technical sphere and in the last
analysis both facilities are equally hard hit. The
economic advantages of nuclear energy, particularly
in the base load range, are thus hardly impaired at
all. To a very large extent it is true that the
lack of orders in the nuclear field is not a
specific feature of nuclear energy. It reflects the
general situation.

A further reason for this reluctance to invest
can be seen from Figure 8 in which the development
over time of the personnel employed in planning and
constructing power stations is shown. This example
involves a coal-fired station and three nuclear
power stations ordered and operated by the Tennessee
Valley Authority (TVA). They were admittedly each
ordered and put into operation at different dates
but they were nevertheless of almost equal block
size in the power category 1 - 1.2 GWe. There is
therefore hardly any direct technical reason for the
considerable rise in personnel with increasing
technical experience. The reason is rather to be
found in the requirements which have been imposed
during the construction period and which in the
course of time have taken on staggering
proportions. Thus, 2575 instructions for technical
modifications to the original design have gradually
accumulated during the construction of Sequoyah.
The maximum of 700 modification requirements was
recorded in 1979, that is to say two years before
the facility was put into operation. This torrent
of requirements must necessarily lead to
construction delays and postponement of the date of
commissioning with corresponding consequences for
the actual plant and electricity generating costs.
At the moment lead times for nuclear power stations
of more than 10 years are being registered in the
USA and this introduces a great uncertainty in any
business, especially in the field of energy supply.

I am explaining this situation for the USA
because I have much data here and the USA has been
the leader in that field. I am not excluding the
fact this situation might also apply elsewhere. In
my own country we have suffered a similar situation
as also in Japan, but in the USA I think it is most
clearly visible. There were strong increases in
capital costs and capital requirements, but looking

Man hours for planning and construction of coal- and nuclear power stations of the Tennessee Valley Authority (TVA) as a function of the start of operation

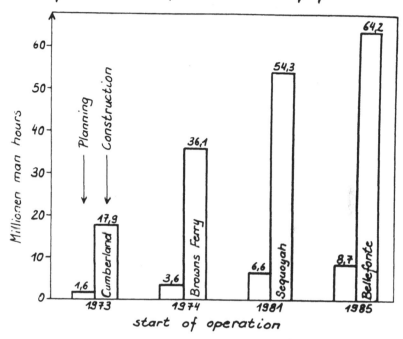

1973: Cumberland, coal fired station
1974: Browns Ferry, nuclear station
1981: Sequoyah, nuclear station
1985: Bellefonte, nuclear station

Source: TVA

Figure 8

around I do not think that this is a typical feature
of the electricity industry, it is rather a general
one. For instance, an initial insight into the
expenditure to be expected for new investments in
the oil and gas industry is provided by Figure 9
which quotes the annual new investments of the
industries in the seventies and extrapolates them up
to the year 2000.

Typical values of specific investment costs
for an output of 1 barrel per day (bbl/d) were in
the Middle East in the range of $2,000. Today
values between $20,000 per bbl/d and $40,000 per
bbl/d are quoted for offshore boreholes and tertiary
extraction measures. This corresponds to an average
of about $400 to $500 per kW of thermal power. If
this value is converted into typical investment
costs comparable to power stations using the
efficiency factor of thermal power stations of 0.33
then we obtain amounts in the range of $1,500/kW,
which is the same order of magnitude as results for
present nuclear power stations. This means that oil
and gas will approximate the situation of
electricity generation when the exploitation of
economically less attractive deposits increases. In
other words, the problem we are considering in the
nuclear field is not really a nuclear problem and
its development essentially brings to an end what I
would like to call, together with Ed Smith from New
York, the first fossil and the first nuclear age.
My thesis is that the first fossil and the first
nuclear age is over.

The first fossil age is characterised by the
statement that oil is clean, unlimited everywhere,
and free. That was the situation before 1973,
because 50 cts/MBTU means essentially that the
energy is free. The typical features for the
nuclear age are that the horizon and the perspective
is only electricity in larger and larger units,
because now a nuclear power station has a block size
of more than 1000 MW (1 GW) electric. In Germany we
have units of 1.3 GWe (the Biblis type) and this is
the result of the first nuclear age, where the units
have been built in ever larger sizes. We should,
however, set question marks with reference to the
product and the advisability of these large units.
In the first nuclear age the statement was that
nuclear energy is as far as one can judge absolutely
safe, the light water reactor would come and later
the fast breeder reactor will come.

If you translate the general disturbance of
the seventies into more explicit notions, then I

Oil and Gas

Annual investments and production in the US.

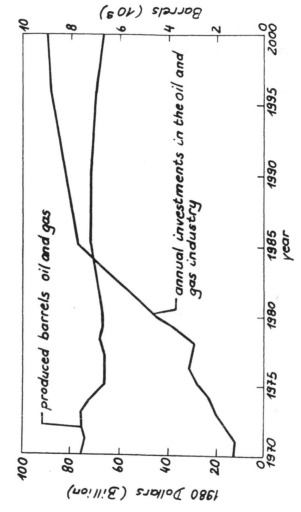

Source: Bankers Trust and EPRI

Figure 9

would say that on the surface much blame was given
to the price of oil. But I think that is only one
part of the story. The increased man hours for
licensing, in the case of nuclear and coal-fired
stations for instance in Germany and the USA, have
nothing to do with the oil price. Underneath the
surface there has been a large environmental
movement and the key words here might be the
"Friends of the Earth" or the "Small is Beautiful"
movement. The claims of the Club of Rome for low
growth, if not zero growth, set the framework - if
you recall the excitement in the early seventies
about the idea of zero growth. Later it was
mandated that you must have zero risk, not small or
acceptable risk or residual risk, but zero risk, and
so we did not get any more licenses for sites.
Therefore siting and obtaining licenses led to very
long leadtimes. The leadtimes of more than ten
years that I mentioned and the large scales of power
plants did not pay off anymore. They turned out to
be more expensive than medium-sized plants so that
600 MW seems to have become a very good size again.

There was, and still is, much talk on
alternatives. Solar power was very much advertised
and, above all, energy conservation, until we
realised that energy conservation is not without
charge. Conservation means that you are
substituting energy by capital and the question is
whether that capital is productive when it is
installed instead of energy. All this resulted in a
very high degree of need for capital.

These are the issues of the seventies that
have led to such turbulences. I think that in the
eighties we will continue to have some turbulence
and so it is perhaps useful to reconsider more long-
range pictures. This only seems to be in
disagreement with Professor Robinson, when he said
that the substitution process for oil is
stochastic. I think there are very surprising
regularities in that substitution process, but they
become visible only in the long run when you
consider the significantly longer time periods.

Some of you might have seen this picture
before, but I will nevertheless show it again
(Figure 10). It shows the fraction that is held by
one primary energy source. Now this is for the
world as a whole, but the figure is typical for a
whole body of information. We have investigated 300
or 400 such examples and therefore it is not just an
incidental plotting. The abscissa starts with the
years 1850 and runs until 2050. A truly long-range

Primary Energy Substitution in the World, 1860 to 2030

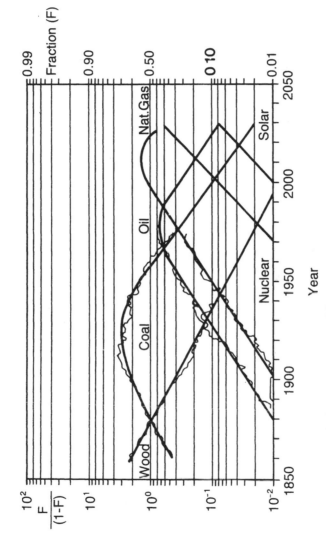

Source: Energy in a Finite World

Figure 10

picture and the remarkable thing is that
regularities exist if you look at the contribution
of each energy source to world energy supply. Wood,
for instance, has left the market. Originally it
held as much as 70% by 1860 when coal entered. Coal
had its peak by 1920, with slightly under 70% of the
market. Then there was the advent of oil and
thereafter the advent of natural gas, but oil is
already peaking and the share of oil in the energy
market is going down whereas in absolute terms it is
increasing. Of course here you see the
fluctuations around the straight line (the thin
lines), but revolutions and world wars have not
changed the secular trends. Up to 1980 the figure
reports on what has happened in the past, while the
others are extrapolations which you may accept or
may not accept. But I find it important to
observe that such regularities, which are really
secular trends, exist. These market pentration
curves are representative of the changes in the
overall energy system not only worldwide but
also more specifically on a national or local
basis. This is extensively documented in our
literature[1].

These changes take time. In terms of energy
you have to consider 30, 40, 50 years and that is
not surprising because 25 years or 30 years is the
natural lifetime of a power plant. Therefore it is
more or less obvious that this is the natural scale
to consider the change of the infra structure
underlying the energy system.

Perhaps more striking is this plot Figure 11)
which considers the H/C ratio in fossil fuels - that
means the hydrogen to carbon atomic ratio in fossil
fuels. Remember that the H/C ratio for wood is in
the neighbourhood of 0.1. The H/C ratio for coal is
somewhere around 1.0. Oil has an H/C ratio of 2
essentially and gas in the form of methane has, as
everbody knows, a ratio of 4. Therefore the past
development of the energy carriers can be
considered as a steady increase of the H/C ratio and
if you take into account this particular kind of
plotting, the straight line is a logistic curve, so
you are considering a transition from an H/C ratio
of zero to an H/C ratio of infinity - that means an
all hydrogen economy in the long run.

Now presently we are are at a ratio of nearly
2 and this enhances the suspicion that in the future
we will have higher H/C ratios and that means that
we will go to gas, away from oil. Looking at the
topics of this meeting there was a talk on oil,

Hydrogen to Carbon Ratio (H/C) of Fossil Fuels in the World, 1860 to 2100

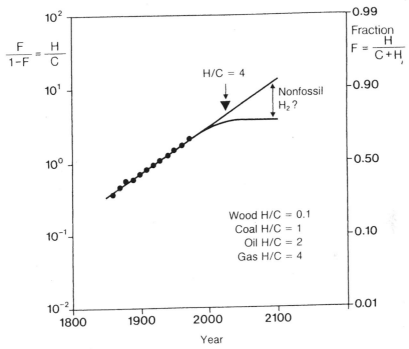

Source. C. Marchetti

Figure 11

there was a talk on coal, there is a talk on nuclear but no talk on gas. I will now try to take the position of people who would conceivably speak on gas, and they probably would say that gas has a bright future. It remains to be seen whether this curve will smoothe out, that is whether we are not going beyond natural gas, or whether that curve continues. If it is to continue, exogenous hydrogen has to be provided. In that case, around the year 2150, the major share of world energy, something like 95% or so, would be based on hydrogen. Indeed many engineers prefer the hydrogen economy and so do I. But not so many engineers are aware that it will probably take 120 to.180 years before we arrive there. Nevertheless, it might be a good way of looking at environmental impacts because the environmental disturbances that go along with the use of fossil fuels might be a limited period in mankind and we are not doomed to do that forever. It gives us a feeling for the timescales at stake. So equipped and with a special view on gas, I would like to confront you with educated guesses for the contributions of primary energy sources to the world primary energy supply in relative and absolute numbers in the long run (Figure 12).

The year 2030 is now no longer the far future to us. So far we have spoken only of the eighties and rightly so if we consider economy. But fast breeder reactors have a different timescale and 2030 is quite a natural date, a benchmark, for considerations of that kind. Now let us discuss the educated guesses, perhaps one might say scenarios, because these numbers are not wildly estimated but based on much scenario work at the International Institute for Applied Systems Analysis and now also at Julich and elsewhere and therefore they are not really arbitrary. I would not be astonished if by the year 2030 we will have a world total demand of 20 billion toe, that is something like 28 terrawatt thermal total and that would mean a little more than a factor of 3 compared with today, because the population will double in that time. Therefore it is not a large figure. How would these 20 billion toe be composed? I think that the absolute amount of oil to be produced will increase to about 4 billion toe and that is not a contradiction of the observation that I made earlier, namely that the share of oil in relative terms might go down to 20%. In the case of gas I personally, together with a few other individuals, expect that the share of gas will increase. The major technical obstacle in

Educated guesses for shares, relative and absolute, in the long run (2030)

	relative	absolute
oil	20 %	$4 \cdot 10^9$ toe
gas	25 % (??)	$5 \cdot 10^9$ toe
coal (solids)	20 %	$4 \cdot 10^9$ toe
nuclear	25 % (?)	$5 \cdot 10^9$ toe
other	10 %	$2 \cdot 10^9$ toe
	100 %	$20 \cdot 10^9$ toe

Figure 12

this point is the long-range transportability of gas. I do not think that one can do it on the basis of liquid natural gas (LNG). The solution might be methanolisation, the methanol route. I would not be surprised if it takes this path, although I shall not be alive to see it. I will have to tell my son to make the observation, that probably 25% of primary energy consumption will be supplied by gas in the year 2030 or the equivalent of 5 billion toe. In the case of coal I am including tar sands and shale oils, further named as solids because all these sources are the ones with the pollution problem, like SO_2, NO_x, etc. The solids would also increase to 4 billion toe or around 20% of the total. Nuclear energy could already hold a 15% share at the end of this century. Therefore an increase to 25% is rather a modest number. The others like wood, biomass and solar power fill the gap to 100%. There may be more reasons for this.

Suppose for a moment such a scenario or such a picture has some element of truth in it. If it tells us that we are enhancing the use of fossil fuels in absolute terms, that we are not abandoning the fossil fuel option, and also, that we are increasing the use of oil and not moving away from it, then we will have to manage the pollution problem today and even more in the long run. Since the clean sources of oil and gas will be used up and exhausted first we will have to go over gradually to fossil fuels which become dirtier and dirtier. I do not think that the availability of resources is the issue; the world has these resources. This observation is also quite different from those made at the beginning of the seventies. Instead, the issue is the cleanliness of their use. That means a different story, a different problem. If we use fossil fuels, which become more and more environmentally unattractive at such a scale, we have to dream of something new. I am not able to predict what this new element might be, but we have our own ideas and there is a club of people which shares these ideas and I will present these ideas to you, not because I know they will materialise but in order to give you a first glance of how these large amounts of fossil fuels can be used without impacting on the environment. That would mean that after the first fossil and nuclear age, we could enter a second fossil and nuclear age with different issues, different problems, different technologies and different features. In that case we have to aim at the more efficient use of fossil fuels in

general. The same holds for new unconventional
resources such as unconventional gas which exists in
very large amounts. All these energies must be used
in combined operation and, therefore, they have to
be integrated into a system with zero emissions to
the environment.

I am speaking as a nuclear engineer who has
been trained to strive for zero emissions. Now let
us see for a moment what happens if you do the same
thing with fossil fuels. We have to close the mass
flows of fossil fuels. It might be intelligent not
to burn the gas directly but to combine it with the
other solid fuel uses in order to get a grip on
environmental effects. This would imply that we
would have to install a second grid. The first grid
is the electricity grid, a closed system. If we are
striving for closed mass flows we must establish a
second grid and the nuclear power would be but one
component of a novel energy system. A novel energy
system would be expensive and the struggle for
capital will be a typical feature of the second
fossil nuclear age. To some extent we have entered
it already if we take into account the systems costs
of at least \$3,000/kW or even more in the case of
oil, as I mentioned at the beginning. The main
problem is the availability of investment capital
for such a second fossil and nuclear age in which
nuclear power would play a role far beyond the
generation of electricity. I personally expect such
developments for the nineties, not in the eighties.
The question is, whether we should prepare for such
a novel energy system or not. Now, what is it after
all?

Let us first have a look at the mass flows
in our current energy system in order to understand
better what should be changed. In our current
energy system (see Figure 13) we are producing
natural gas, we are bringing it in a pipe to the
consumer, and the consumer produces CO_2, H_2O and
NO_x, and the NO_x produces acid precipitation.
Crude oil is first given to the refinery and then
delivered to the consumer and the refinery
produces CO_2, H_2O, NO_x, and SO_2. SO_2 is another
component of the not so well understood phenomenon
of acid precipitation. But that it plays a major
role, is without question.

Besides steel production, coal is normally
directly fed into electricity generation today so
that the consumer using electricity causes no
environmental pollution and all the blame goes to
the power station with its release of CO_2, H_2O,

NO_x and SO_2 into the environment. Other efficient
producers of electricity are the nuclear power
stations, mainly light water reactors, which deliver
their product directly into the first grid. While
fossil power stations convert their fossil energy
sources directly into electricity, nuclear power
requires a step before the power station in which
the fuel is conditioned. This involves chemical
handling of the yellow cake of the uranium ore U_3O_8
and enrichment and this in itself is already a large
industry in the case of nuclear energy.

Now with these observations let us consider
what a second generation, a second age for fossil
and nuclear power, could possibly look like. In
Figure 14 you see that we have three stages, the
consumer, the power station and the fuel conditioning, but now for all primary energy sources and not
only for nuclear power. The fuel conditioning is
based on the idea of decomposing any energy input
and of cleaning it thereby as early as possible from
impurities in order to have clean intermediate fuels
that will be transferred to the power stations via a
second grid. Let us go into this in greater detail.

We first start with the decomposition of air.
This is effected by the separation of nitrogen from
oxygen and will reduce the NO_x emissions of
today's open cycle combustion with air to about one
third. For reasons of flexibility it would be
useful to withdraw the oxygen not only by air
separation but also by water splitting which is
effected by means of electrolysis in our example.
But the decisive step in this new supply structure
is the treatment of solid fossil fuels, which
include both coal and, in principle, all kinds of
crude oil. The molten iron bath process with its
1400 °C, which breaks any chemistry and purifies the
fuel from the impurities, is given here as an
example of the treatment of these solid fuels. From
this process we get the useful and clean substances
H_2 and CO, whereas the sulphur and all the other
impurities are left in this process in a depositable
form. Finally, natural gas will not be burned
directly anymore. Instead, it would undergo a shift
reaction together with water, known as steam
reformation, and would result in H_2 and CO. In
refining the fossil fuels we first of all obtain the
intermediate energy sources of oxygen, hydrogen and
carbon monoxide, which then provide electricity and
methanol or synthesis gas via a conversion stage.
These final energy sources can be largely used in an
environmentally safe manner for we are then only

left with CO_2 for which we still have to find possibilities of application. And only in the case of methanol end uses is there still some NO_x to be expected, but in significantly smaller amounts.

In this novel fossil energy system nuclear energy will primarily be utilized in the electricity sector and will be increasingly supported by the breeder. The breeder's task will not be confined to generating electricity but it will also supply suitable fuel for the reactor systems still utilized, the LWR, which would be supplemented by the HTR. In comparison to all other reactor systems the HTR has the special advantage of a high coolant gas temperature and can thus provide the high process steam temperatures necessary for methane splitting and coal refining in one operation. The HTR creates a direct connection between the nuclear and fossil option by generating process heat at more than 950 ^{o}C as a major product for fossil refining and electricity as a by-product, and thus initiates through steam reformation the entry of nuclear energy into the energy market beyond electricity.

The decisive advantage of this new energy system is to be found on the producer side. Since the intermediate energy sources, like O_2, H_2, and CO, are supplied in parallel from other sources, we can achieve the urgently required flexibility on the producer side exclusively with technologies already familiar to us. This guarantees a crisis-proof energy supply. Remember that the main problem of our actual supply system is that the consumer and not the producer has to perform the fuel substitution, mainly from oil to gas or coal. That is why we at the KFA-Julich are working on this novel energy system. I am not saying that this novel energy system will be realized that way, but I am saying that my own laboratory is working very intensively along these lines, the energy laboratory at the Massachusetts Institute of Technology (MIT) is doing it, the city of Stockholm is strongly engaged in doing it, and additional partners are envisaged.

I personally would not be surprised if this novel energy system, in terms of systems cost, would end up to be cheaper and not more expensive. This applies especially if you take into account the costs for the reduction of emissions caused by combustion of large amounts of dirty fossil fuels in the future. Therefore, I think we had better make up our minds and reflect on what a second fossil and nuclear age would possibly look like.

Mass Flows In Current Energy Systems

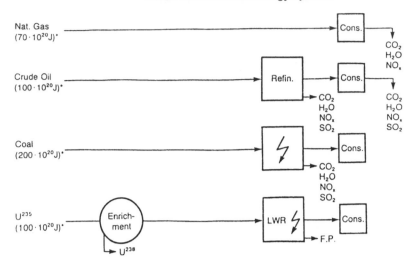

* these numbers reflect perception of resources as seen until the early seventies

Figure 13

Mass Flows in Novel Fossil Energy Systems

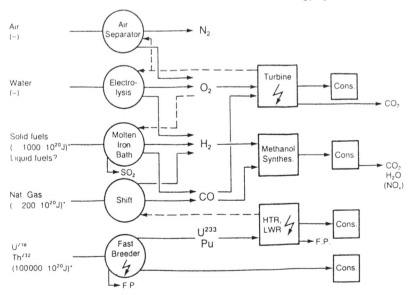

* these numbers reflect perception of resources as seen since the early eighties

Figure 14

The Future of Nuclear Energy

REFERENCES

1.   "Energy in a Finite World - A Global
Systems Analysis" Report by the Energy Systems Group
of the International Institute for Applied Systems
Analysis, Wolf Häfele (Program Leader). Cambridge,
Massachusetts, USA, Ballinger Publishing Company.
March 1981.

ACKNOWLEDGEMENTS

The author is very much indebted to Dr Wolfgang Jaek
and Miss Jenifer Hiscox for transforming the
verbatim transcript into this written version.

OECD COAL DEMAND IN THE 1980s

Herman T Franssen

1. INTRODUCTION

Coal played a dominant role in world energy use in the early part of this century. As late as the early Fifties when the transition to oil and gas began in earnest, coal still contributed 57% to OECD primary energy consumption. Not only was coal the principal fuel in electricity generation, as it still is in the U.S. today, but it was widely used in other sectors as well, as shown in Table 8.

Table 8

Distribution of OECD Coal Use by
Major End-Users in 1950 and 1980

(in %)

	1950	1980
Electricity	19	66
Coke Ovens	20	22
Industry	22	0.5
Railroads	12	negl.
Residential/Commercial	22	3
Conversion	5	0.5

The rapid shift to clean oil and gas in the 1950s and 1960s quickly reduced coal's role in all sectors but electricity and the iron and steel industry. In volumetric terms coal consumption remained stagnant in the 1960s and in fact through the late 1970s. Coal's share of OECD energy markets however, declined from 36% of Total Primary Energy (TPE) use in 1960 to 20% in 1973. Availability of cheap and cleaner alternative to oil, handling problems, etc., all contributed to the shift away from coal. As shown in Table 1 the big decline in coal use was largely caused by declining demand for thermal use by non-electricity industries, heating use in the residential sector, and the shift of the railways to oil and electricity. This was partly offset by growth in the electricity sector, mainly in the United States.

Continuing decline in the share of coal in TPE in the 1960s caused energy analysts to state unequivocally that coal had no future in OECD energy matters. It was a dirty fuel, difficult to handle, non-competitive with fuel oil and sometimes natural gas, and required cumbersome processing and transportation to markets.

The oil shock of 1973 and more so the supply disruptions in 1979 followed by another round of price increases, changed the perceptions of the future of coal use. In the late 1970s when the perception of structural changes in energy demand had not yet become apparent, it was assumed that as quickly as possible all sources of energy in the OECD needed to be mobilised in order to reduce over-reliance on imported oil. The OECD Steam Coal report of 1978 and the famous M.I.T. Wocal study reflected the renewed optimism for future coal use. This optimism was again expressed by IEA ministers at the Venice Summit Conference of 1979, when Governments showed their determination to see to it that coal use in the IEA Countries would be doubled in the 1980s. The lengthy post-1979 recessions and possible major structural changes in heavy industries of the industrial countries have recently tempered the atmosphere of buoyant optimism on coal use of the late 1970s. Current assessments of the use of coal in the next two decades still show great promise for the "black gold" in the OECD's future energy mix, but some of the optimism of the late 1970s has vanished. This is the result of lower economic growth expectations, expected changes in the mix of industrial output, and structural changes in energy markets.

2. COAL IN THE 1960s
Between 1960 and 1973, demand for coal remained virtually constant in the OECD area, increasing from 964 Mtce in 1960 to only 998 Mtce by 1973 (see Table 9) despite the rapid rise in TPE demand. The relative importance of coal accordingly declined from more than 36% of TPE in 1960 to less than 20% by 1973. This overall performance is explained largely by greatly reduced coal demand for thermal use by non-electricity industries and heating use in the residential sector, primarily in Europe, which

TABLE 9

OECD COAL USE BY SECTOR
(Mtce)

	1960				1973				1980			
	Electricity	Coke Oven	Other	TOTAL	Electricity	Coke Oven	Other	TOTAL	Electricity	Coke Oven	Other	TOTAL
North America	169	79	99	347	334	93	60	487	473	47	76	596
Pacific	31	18	47	96	34	72	26	132	68	48	22	138
Europe	126	134	261	521	169	118	92	379	259	55	85	399
OECD Total	326	231	407	964	537	283	162	998	800	150	183	1133

Source: OECD Energy Balances

Memorandum Item

	1960	1973	1980
1. Share of Coal in Total Primary Energy (%)	36.3	19.5	21.7
2. Share of Coal in Total Fuel Inputs for Electricity Generation (%)			
North America	47.0	41.9	48.0
Pacific	48.9	18.3	23.4
Europe	53.0	37.5	40.1
OECD Total	49.4	37.4	42.3

which was partly offset by growth in the electricity
sector, mainly in the United States. Reflecting the
steady growth of the iron and steel industry in the
OECD in the 1960s, metallurgical coal use for coke
ovens showed a modest increase of 1.6% a year from
231 Mtce to 283 Mtce.

Within this overall picture, there are
considerable regional differences. Coal use in the
electricity sector in North America virtually
doubled, while growing only slightly in other
regions. As a result, North America accounted for
almost 50% of total OECD coal use by 1973, compared
with only 36% in 1960. Coal use in the iron and
steel industry declined slightly in Europe, but
quadrupled in the Pacific area. Virtually all of
this was due to the fast growth of the Japanese iron
and steel industry. In other uses, coal declined in
all OECD areas, but most dramatically in Europe.

Even in the years following the sharp oil
price increases of 1973, total OECD coal demand did
not grow significantly. Between 1973 and 1978, coal
demand grew only at a low rate of 0.9% annually and
the share of coal in TPE decreased slightly from
19.5% in 1973 to 19.2% in 1978. During this period,
the demand growth of coal for electricity generation
in each OECD region was partly offset by the decline
in coal use for coke ovens in North America and
Europe, and for the residential sector in Europe.
However, the oil price increases following the
Iranian revolution caused a mini-boom for coal in
1980 and the share of coal in TPE increased to 22%
in 1980. Despite the slow overall growth in total
demand, coal consumption for electricity generation
increased at a fairly high rate of 5.1% per year
since 1973, thus firmly establishing electricity
generation as the dominant market for coal. It
accounted for 66% of total coal use in 1980 compared
with 55% in 1973. The considerable volumetric
growth in coal use for electricity generation did
not, however, prevent a substantial decline in the
share of coal in total OECD fuel inputs for
electricity generation, from almost 50% in 1960 to
about 37% by 1973. In particular in Western Europe
and Japan the relative importance of coal in this
sector declined dramatically because of the
availability of the cheaper and cleaner fuels, oil
and natural gas. Even after the 1973/74 oil shock
there was no significant growth in coal demand.
Substantial growth in coal demand in the electricity
sector (about 5%/year) was largely offset by
declines in the industrial and residential/

commercial sectors.

The continuous, and probably irreversible, decline in coal use in the residential sector is mainly due to the inconvenience of handling coal and the environmental problems associated with it. As for metallurgical coal, the general decline in steel output in the OECD caused demand for metallurgical coal to fall in the years after 1973. In the industrial sector, many boilers using thermal coal had been replaced by oil or gas boilers by 1973 due to the cost advantage, convenience and environmental attraction of oil and gas. Between 1973 and 1979 no big change in thermal coal use in this sector took place, but following the 1979-1980 oil price rise the cement industry switched its fuel from oil to thermal coal. There were also signs in some other energy-consuming industries of switching from oil to coal.

3. COAL PRICES

The distinct difference in price trends between coal and oil has been the main factor responsible for explaining the relative use of these fuels in the industrial sector and particularly for electricity generation. During 1960 and 1973, coal prices moved more slowly than oil in most of the coal consuming and producing countries, except in the United States (see Table 10). Among consuming countries, coal prices in Japan declined by about one-third between 1960 and 1973, due largely to replacement of high cost domestic coal by imported metallurgical coal. At the same time, the price of heavy fuel oil progressively declined in real terms, particularly in Japan, Italy and France. These countries had no significant domestic coal resources and consequently experienced a rapid transition from coal to oil in their industrial and residential sectors. In addition to the incentive provided by relative price movements, environmental considerations and ease of handling also promoted the shift away from coal, particularly given the less sophisticated technology available to control emissions from coal burning in the 1960s.

Following the considerable increase in oil prices after 1973, coal prices also tended to rise in all OECD countries. However the rate of increase of coal prices was markedly lower than that of oil. Coal price rise was most prominent in Germany, where coal production was increasingly made at more costly deeper seams. In other countries, however, between 1973 and 1980, real coal prices rose by only 4%-8%

TABLE 10

HISTORICAL DELIVERED COAL AND OIL PRICES FOR INDUSTRY
(Real Price Indices, 1973 = 100)

| | 1960 | 1973 | 1980 | Annual Average Increase (%) | |
				1960-1973	1973-1980
United States					
Coal	62.2	100.0	141.6	3.7	5.1
Heavy Oil	82.0	100.0	291.9	1.5	16.5
Japan					
Coal	145.6	100.0	143.7	-2.8	5.3
Heavy Oil	177.0	100.0	278.0	-4.3	15.7
France					
Coal	84.3	100.0	167.4	1.3	7.6
Heavy Oil	136.1	100.0	291.0	-2.3	16.5
Germany					
Coal	82.7	100.0	210.7	1.5	11.2
Heavy Oil	106.4	100.0	204.9	-0.5	10.8
Italy					
Coal	87.7	100.0	166.0	1.0	7.5
Heavy Oil	157.7	100.0	218.0	-3.4	11.8
United Kingdom					
Coal	99.8	100.0	129.0	0.0	3.7
Heavy Oil	106.7	100.0	222.8	-0.5	12.1
Canada					
Coal	124.3	100.0	129.6	-1.7	3.8
Heavy Oil	128.0	100.0	254.4	-1.9	14.3

Note: Coal prices for industrial use include both
         thermal and metallurgical coal.

Source: Government statistics and other sources.

per year compared with an 11%-17% annual increase in
real heavy fuel oil prices for industrial use. In
the United States, coal prices rose by about 40%
from 1973-80, while heavy fuel oil prices almost
tripled over the same period. In most other OECD
countries, coal prices increased much more modestly
compared with industrial heavy fuel oil prices. As
a result, coal's competitive advantage increased
dramatically after 1973, and this has been even
further accentuated by the most recent increases in
oil prices in 1979 and 1980.

   The price of coal increased rapidly during
1980 and 1981 due to the tight market caused by
the sudden decrease of Polish exports to OECD Europe
and the rapid conversion of the cement industry to
coal.

4. 1979-82 DEVELOPMENTS

In the first two years following the second oil
shock of 1979 coal demand grew rapidly in spite of
generally declining energy consumption. Total
primary energy consumption fell 2.5% per year and
oil demand was reduced by 7.5% per year through late
1981. By contrast OECD Coal Consumption grew by 17%
in those two years. Coal demand which had remained
stagnant in the 1960s and early 1970s, and grown by
only 30 Mtoe in the five years after the first oil
shock, increased by 100 Mtoe between 1979 and 1981.
Growth rates between 1979 and 1981 differed by
region and country, ranging from 12% in Western
Europe to 25% in OECD Pacific.

   In the US the growth was largely concentrated
in the electricity sector, while in Japan industrial
coal use grew most rapidly. In all OECD regions the
cement and iron and steel industries converted
quickly to coal; in Japan the paper and pulp
industry also moved rapidly out of fuel oil and into
coal. The rate at which these changes took place
explains the high overall rate of growth of coal
demand. For example, only 16% of the cement
industry in Japan used coal in 1979. By 1982 about
95% of all cement was produced with coal as the
principal fuel. Changes from fuel oil to coal in
the North American and European cement industries
were also significant, but not as impressive as in
the Japanese case.

   Contrary to expectations, overall coal
consumption declined again in 1982 (about 2.3%), but
the rate of decline was smaller than the reduction
in total primary energy demand (3.5%). As shown in
Table 11 almost the entire reduction in coal use was

in the United States.   The absolute level of coal
use in Western Europe and Japan remained practically
constant.   Most of the fall in US coal demand was
due to reduced consumption of coking coal and
steam coal in the industrial sector.   While coal in
the US electric utilities was down only 0.5%, coking
coal demand was down 23% and industrial coal use
4.5%.

Table 1.1
OECD Coal Use in 1982

(in Mtoe)

	1981	1982
OECD total	825.8	806.7
North America	435.8	413.0
Pacific	104.3	108.9
Western Europe	285.7	284.8
Canada	27.0	29.4
US	408.9	383.7
Japan	69.1	68.2
France	32.7	30.5
West Germany	86.4	86.0
Italy	14.4	14.7
UK	67.8	63.8

The price of internationally traded coal
increased rapidly in 1980/81 due to the tight market
caused by the sudden decrease of Polish imports and
increased conversion to coal.   Since the beginning
of 1982, however, world coal prices declined somewhat due to lower demand under depressed economic
conditions combined with normal activities by all
coal exporters who were not affected by any notice-
able industrial action in that year.   The most
important event on the supply side was the return of
Polish output to the 1980 level.   The combination of
the above events on the demand and supply side
caused a fall in US and South African imports to
Western Europe.
    In the early part of 1983 coal demand did not
show any signs of growth, causing further downward
pressure on prices.   Softness in the market has
resulted in traditional importers reducing the
offtakes from their long-term contracts to a minimum
in order to take advantage of current low spot
prices.
    Average US and Australian FOB contractual
steam coal prices in 1982 ranged between $43 and $54
per metric ton.   In recent months FOB prices of $37

- $45/metric ton have been quoted in South Africa, Australia and the United States. South Africa, Australian and US steam coal cargoes are landing in Europe at CIF prices between $45 and $55/metric ton, or some 20% - 30% lower than average CIF prices last year. Polish coal is offered systematically in western Europe at CIF prices often well below the CIF prices of steam coal from the US and South Africa. The above figures are purely indicative, as they only relate to marginal quantities. However, these quotations do reflect current trends in price negotiations for long term contracts.

Soft market conditions are likely to persist for some time, given continuing low economic activity combined with high stock levels in many parts of the coal chain and continuing reduced steel production in OECD countries.

5. COAL DEMAND IN THE 1980s

In spite of the current slack in the coal markets, the long term prospects for coal in the OECD countries are good.

Coal can compete with other fuels either in the form of direct burning or coke, converted to gases, liquids and electricity, or as a mixture with other fuels. Because coal reserves are readily available in the OECD countries in vast quantities and are now cost-competitive compared with liquid fuels, substantial growth in coal demand is expected to take place. After 1973, various steps have been taken to encourage coal use, production and trade, and the first positive results became apparent in 1979 when the use of coal in OECD countries reached a record level of more than 1,000 million tons of coal equivalent (1090 Mtce). This trend continued and the demand increased to 1133 Mtoe in 1980.

Earlier studies after 1973 have shown the rapid increase in coal use (see Table 12). The study by the IEA in 1978 (Steam Coal - Prospects to 2000) projected that coal use will reach about 1,650 Mtce in 1990 and around 2,000 Mtce by the end of the century. It assumed high penetration of nuclear power in the electricity sector in 2000, reflecting the more favourable atmosphere surrounding nuclear at that time. Further, the increase in coal use in the industrial sector was projected to be modest, as the price differential between oil and coal was not big enough for industrialists to switch from oil to coal. Consequently, the projected coal use in 2000 was smaller compared with other studies. Following the price rise of oil in 1979 and in 1980, the cost

TABLE 12

EARLIER PROJECTIONS IN OECD COAL DEMAND
(Mtce)

	1985	1990	2000
1. Steam Coal Prospects, 1978, IEA			
Projected Coal Use	1407	1632–1688	1879–2106
(Electricity)	(934)	(1190–1146)	(1211–1438)
(Metallurgical)	(319)	(342)	(378)
(Others)	(154)	(200)	(290)
2. WOCOL, 1980			
Projected Coal Use	1235–1370		2000–3025
(Electricity)	(800–880)		(1325–1850)
(Metallurgical)	(285–295)		(330–375)
(Others)	(150–195)		(345–800)
3. IEA Reference Case in 1981			
Projected Coal Use		1719	2648
(Electricity)		(1023)	(1584)
(Others including Metallurgical Coal)		(696)	(1064)
4. IEA World Energy Outlook, 1982			
Projected Coal Use	1237	1500–1546	2129–2486
(Electricity)	794	(983)	(681–1667)
(Other)	443	(563–517)	(819–739)

advantage of coal compared with liquid fuels has been greatly reinforced. At the same time, the projections on nuclear power have been continuously revised downward due to the increased public objections etc., and coal has been expected to increase its share in the electricity generation sector. Reflecting these developments surrounding coal, the WOCOL study in 1980 projected that coal use could reach 3,000 Mtce in 2000. The expanded coal use was also assumed in the IEA reference case in 1981. The case projected 50% increase in coal use in the 1980s and another 50% increase toward the end of the century from the 1990 level.

The optimism for coal use expansion, which is represented by the WOCOL study, has been somewhat fading since 1981. Oil prices have been suffering the downward pressure by the current supply surplus in the oil market. At the same time, various problems in converting away from oil to coal have become clearer as is explained later. For these reasons, the recent arguments for coal demand in the 1980s tend to show smaller figures than those which prevailed a couple of years ago.

Table 13 shows current estimates of OECD coal demand in the 1980s, based on the 1982 government submissions to the IEA. The table shows further downward revision from the 1981 IEA reference case, but still projects a substantial increase in coal demand. Here, coal demand is projected to increase 30% to 1646 Mtoe in 1990 from 1133 Mtce in 1980. However, future coal use in utilities and in industries could be further delayed due to the depressed electricity and energy demand in general, and the possible substantial decline in real crude oil prices. Considering these circumstances, the figure in Table 6 could still be optimistic. It can be said, however, that this is the potential coal demand at this particular time in the 1980s, assuming governments continue to encourage coal use expansion. To grasp the current situation correctly, analyses of coal demand must be done sector by sector from both sides, positive and negative.

5.1 Electricity Sector

Table 13 shows that coal demand in the electricity sector is expected to grow 25% in the next ten years at the annual increase rate of 2.3%.

About two-thirds of coal is currently used in electricity generation. Because the dominant portion of coal use is projected to remain in the

TABLE 13

OECD POTENTIAL COAL USE BY SECTOR
(Mtce)

	1981		1985		1990	
Electricity	Mtce	% Share[4]	Mtce	%Share	Mtce	%Share[4]
OECD Total[1]	778	39.3	822	38.5	1006	39.4
North America	487	47.7	520	46.7	623	46.9
Pacific	53	18.2	74	21.7	103	24.1
Europe	237	36.3	228	33.0	280	34.6
Industry[2]						
OECD Total	268	19.7	419	24.0	485	24.7
North America	110	17.0	214	23.6	240	23.7
Pacific	66	28.2	83	30.7	99	31.5
Europe	92	19.0	122	21.4	146	23.0
Other[3]						
OECD Total	111		125		155	
Total OECD Coal Use	1157		1366		1646	

1 Figures do not necessarily add due to rounding.

2 Includes coking coal for iron and steel industry.

3 Includes household sector, synfuels feedstock and other
   solid fuels.

4 For electricity generation, this represents the share of
   total electricity generated by solid fuels. For
   industry, this represents solid fuel's share in total
   inputs.

TABLE 14

ELECTRICITY GENERATION CAPACITY BY FUEL SOURCE IN 1980 (GW)

	Coal	Oil	Gas	Nuclear	Hydro/Geo	Total
United States	249.0	148.0	77.0	55.0	81.0	610.0
Canada	15.0	8.3	4.5	5.6	47.9	81.3
North America Total	264.0	156.3	81.5	60.6	128.9	691.3
Japan	–	84.7	–	15.5	29.0	129.2
Australia	14.6	1.7	1.7	–	6.1	24.1
New Zealand	0.2	0.7	0.8	–	4.0	5.7
Pacific Total	14.8	87.1	2.5	15.5	39.1	159.0
United Kingdom	47.0	16.8	–	6.5	1.4	71.7
Germany	43.3	14.7	13.6	8.6	6.5	86.7
France	14.3	15.8	1.7	12.8	19.3	63.9
Italy	4.4	24.4	–	1.4	16.2	46.4
Spain	5.3	11.1	–	1.1	13.3	30.8
Others	8.4	23.6	1.7	14.0	61.3	109.0
OECD Europe Total	122.7	106.4	17.0	44.4	118.0	408.5
OECD Total	401.5	349.8	101.0	120.5	286.0	1258.8

Source: IEA Estimates

electricity generation sector in the future, demand growth of electricity becomes a critical factor in determining the expansion of coal use. Although electricity demand is expected to grow constantly in the long term because of its cleanliness and convenience, there are some signs that demand growth of electricity might be low in the 1980s compared to the 1960s and 1970s when electric demand increased by more than 5% per year in the OECD. The economic growth of the early part of the 1980s is under successive downward revisions. The relation between the growths of electricity demand and economy is also uncertain. Many countries have begun to think that demand growth of electricity is certainly lower than the economic growth in the 1980s. It is not known exactly what was happening in each end use level to explain the low demand for electricity in 1980 and 1981, but at least we can guess that the economy is under the adjustment stage to high energy costs and this trend might continue in the 1980s.

Another critical factor is the penetration of coal in generation fuel mix. Table 14 shows the share of each fuel in electricity generation. The figures are again based on the 1982 governmental submissions and correspond to the coal use in Table 2. The share of coal is projected to decline somewhat from the current 42% to 39% in 1990, as the dominant portion of the incremental electricity demand is projected to be absorbed by the rapid expansion of nuclear power. Here, the comparative costs of generation among fuels become the determining factor in the long run. IEA assessments, as presented in the World Energy Outlook (1982), suggest that with the exception of North America, the generation cost of nuclear power plants are generally lower than that of coal-based power plants by about 20% in Europe. Assuming interest rates of 10% and a "normal" construction leadtime of 6 years nuclear power is also cheaper in Japan by about 10%. However, nuclear power is very vulnerable to high interest rates and longer leadtimes, as financial costs snowball rapidly due to the big capital investments required.

The current high interest rates in North America and in Europe, and the lengthening leadtime of nuclear power plants are eroding the financial advantage of nuclear power. Actually, there has been only a few new orders for nuclear reactors in the last three years and numerous cancellations in the US. Thus, in the 1990s, coal may be expected to absorb much of the incremental demand for

117

electricity generation. The projected expansion of nuclear power in the 1980s is partly due to the fact that many nuclear plants are already under construction and that substantial financial commitment has been made already. However, recent developments in the State of Washington in the US have shown that partial completion and financial commitments do not exclude cancellation.

Table 15

Projected Fuel Input in Electricity
Generation (%) in the OECD

	1980	1985	1990
Coal	42	38	39
Oil	16	13	8
Gas	10	8	8
Nuclear	12	21	27
Hydro/Others	20	20	18
TOTAL	100	100	100

Oil prices also affect the demand for coal in the electricity sector, as they determine the rate and the extent to which existing oil plants will be converted away from oil to coal. Table 14 shows the installed capacity of electricity generation by fuel source. In 1980, oil-based power plants still accounted for 28% of the total generation capacity. One-third of these oil plants were estimated to be either phased out or converted to coal plants in the next ten years, assuming the crude oil price of 34 dollars per barrel. However, crude prices have now been reduced by about $5 per barrel and may remain at this level for some years. Lower prices are likely to result in postponement of conversion plans.

The importance of coal in OECD electricity generation differs widely between countries as shown in Table 14. Countries where coal accounts for a major share in electricity generation are the traditional big producers: the United States, Australia, Germany and the United Kingdom. Among these, the United States and Australia will increase coal use in power plants substantially as their low cost domestic coal has economic advantage over other alternatives.

In the United States, the major thrust of coal is expected in the 1990s, as nuclear seems to have lost economic advantage against coal due to high interest rates and long leadtimes. However, demand

118

growth in the 1980s is still uncertain due to the
anticipated lower electricity demand and to the
uncertainty in oil prices. Another factor hindering
the rapid conversion to coal in the United States is
a regulatory matter. State regulations on
electricity pricing have left many utilities with
financial difficulties. A utility must be
financially strong to engage in the expensive
conversion of existing capacity. Thus, the
regulatory climate can be a major impediment to
conversion of many existing oil and gas-fired
electricity plants to coal. Financial constraints
in part due to regulatory problems may prevent
utilities from building new plants or converting
existing ones in many parts of the US. This
accumulation of delays and cancellations together
with the linkage between economic growth and
electricity usage may result in oil and gas use
remaining above economic levels.

In OECD Europe, coal use in power plants is
estimated to grow as a whole though projected growth
varies greatly among countries. Italy and Spain
have ambitious coal projects, but insufficient
infrastructure could delay the massive expansion of
coal use. France foresees a substantial drop in
this sector as nuclear will gradually replace many
existing base load coal-fired plants.

In Japan, electric utilities are planning to
diversify fuel sources away from oil to a
combination of coal, nuclear and LNG, and the use of
all these fuels is estimated to grow rapidly. One
new coal-fired power plant (1GW) has already been
completed and several others are either under
construction or in the planning stage. Moreover,
several oil-fired power plants are planned to be
converted to coal. It is note-worthy in Japan,
however, that an LNG power plant can compete with a
coal-based power plant despite its high fuel costs
which are linked to crude oil prices. Due to the
strict regulations aginst SOx, LNG power plants have
become the only choice among fossil fuels to be
built close to population centres. High fuel costs
of LNG, thus, can be offset by lower costs of
sending electricity from power plants to end users.

In the OECD as a whole, coal use in
electricity generation will increase substantially
toward the end of the century. However, to what
extent it will increase in the 1980s is still
uncertain, considering the various factors analysed
above.

5.2   Demand of Coal for Coke Ovens

Demand for metallurgical coal for coke ovens has
tended to fluctuate around a constant level of 250
million tons in the last two decades, although crude
steel production nearly doubled during this period
(see Table 16). This was largely due to the
modernization of steel production and the growing
use of fuel oil as asupplemental heat source. As a
result, the specific consumption of coke per ton of
crude steel has decreased. During the early 1980s,
a large part of the oil and gas consumption in the
iron and steel industry (about 50 Mtoe in 1979) will
gradually be phased out and this will offset the
decline in coal use by modernization. Growth of
crude steel out-puts, moreover, becomes a critical
factor in determining coal demand for coke ovens.

Based on government estimates, demand for coal
for coke ovens is projected to grow 36% in the
1980s at an annual rate of 2.7%, from 241 Mtce in
1980 to 313 Mtce in 1990. Crude steel production
has stagnated since 1975 despite the growth, though
at a low rate, of the economy. This trend of crude
steel production (i.e. no growth) will not last for
ever, but it can reasonably be assumed that the
growth of crude steel production would be
substantially smaller than the economic growth.
Assuming OECD economic growth rates of 3.0% annually
in the 1980s, growth of crude steel output may be at
best 70% of economic growth, or 2% per year. If
this is the case, the demand for coking coal would
be at most 2.0% annually. Some Member countries
have ambitious plans for steel production, but this
could turn out to be optimistic.

5.3   Industrial Sector

In the OECD as a whole, about 8 Mbd of oil is used
in the industrial sector compared with 4.7 Mbd in
power generation. Most of the energy used in the
industrial sector is for process steam, direct heat
and space and water heating. These are the areas
where coal could gradually replace oil. Coal use in
the industrial sector was about 100 Mtoe in 1980. A
study by the Coal Industry Advisory Board of the IEA
suggests that the technical potential for converting
from oil to coal in industry is significantly
greater than had previously been thought and that
steam coal use in the industrial sector could
potentially reach as much as 700 Mtce. However, the
full technical potential is not likely to be reached
because of financial, environmental and other
difficulties in converting from oil to coal.

120

TABLE 16

OECD CRUDE STEEL PRODUCTION AND COAL USE IN COKE OVENS

	Crude Steel Production (Million Tons)	Coal Use in Coke Ovens (Mtce.)
1961	235	226
1962	234	222
1963	252	226
1964	293	244
1965	305	253
1966	311	253
1967	325	250
1968	348	259
1969	383	275
1970	392	288
1971	368	266
1972	403	266
1973	457	283
1974	458	287
1975	384	273
1976	409	269
1977	393	246
1978	413	230
1979	433	247
1980	398	241

The immediate potential for fuel switching
lies in the energy-intensive industries that require
bulk heat or a combination of heat and power. Coal
use is particularly favoured in the cement industry
where ash and sulphur become part of the product,
and possibly at paper mills where the configuration
of wood-burning boilers is suitable for coal. In
addition, great opportunities for bulk steam-raising
exist in the aluminium, textiles, chemical, food and
engineering industries. Coal could also provide
energy to oil refineries through the direct use of
coal (or coal-oil mixture) in the main boiler, and
through gasified form for steam and process heat,
replacing oil and gas-fired boilers.

Despite the market potential in this sector,
however, there are some reasons for concern as to
whether coal could displace oil on a massive scale
during the 1980s. Initially, the iron and steel and
cement industries switched quickly back from oil to
coal, but most plants have now already been
converted and future demand growth for coal in these
sectors will depend primarily on growth in output.
Other industries have various obstacles to switching
from oil to coal. These include existing boiler
stocks, industrialists' criteria for fuel switching,
inadequate infrastructure and environmental
concerns.

5.3.1 Existing Boiler Stocks

Two decades of readily available and inexpensive oil
and gas have left the industry sector heavily
dependent on hydrocarbons. The existing boiler
stock is dominated by oil and gas boilers installed
over the last twenty years. The two main types are
either fire-tube or water-tube boilers with the
latter forming 75% of fuel use. The life expectancy
of water-tube boilers is 40 to 50 years, and hence
boilers do not need replacing for 20 to 40 years.
In France, for instance, 60% of boiler fuel in 1975
was oil and gas burnt in boilers constructed in the
previous 10 years. Rapid conversion can thus only
be achieved through early retirement, which would
require additional investment. With the exception
of the iron and steel and cement industries,
switching to coal has not yet become widespread and
is unlikely to take place to a major extent in the
near future without supportive policy frameworks,
even in energy-intensive industries such as pulp and
paper, aluminium, glass, textiles and sugar, where
immediate potential for expanded coal use exists.

The pulp and paper industry depends heavily on

petroleum fuels and would profit by switching to
less costly energy sources. Boilers are usually
installed to burn mill-generated wastes and
purchased wood in combination with fossil fuels.
The capital cost of wood-coal combination boilers is
not much greater than that of wood-oil combination
boilers. Further, the configuation of wood-coal
combination boilers is similar to that of boilers
constructed to burn either fuel separately.
Therefore, compared with other industries, the pulp
and paper industry has lower capital cost hurdles
standing in the way of increasing coal use. The
industry in Japan is experiencing conversion now and
more than half of the oil boilers in the industry
have been converted in recent years. In other
areas, however, the initial response of the industry
to the oil price hike has been to expand use of
biomass fuels from wood residues and surplus timber
that are available in many regions that produce pulp
and paper products. In addition to burning mill
residues, mills in North America, Brazil and other
wood surplus regions are likely to burn surplus non-
marketable wood from the forests. Outside the
world's heavily wooded regions, especially in
central and southern Europe, mills tend to be small
and dispersed and many of them have been operating
at low profit levels. It is therefore unlikely that
early retirement of existing boilers, requiring
extensive investment, will take place. For these
reasons, switching to coal is likely to be slow in
the pulp and paper industry.
    The aluminium industry uses electrical energy
and the majority of existing smelters operate with
locally available fuel sources such as hydroelectric
power and natural gas for electricity generation.
Even in the case of plants using oil for power
generation, changing boilers from oil to coal
requires a large investment and generating plants
are often not big enough to use coal economically.
It is unlikely that existing smelters will convert
to using self-generated, coal-fired electricity. As
for the future demand of aluminium, it is generally
agreed that growth of aluminium consumption will
exceed general economic growth. There is, therefore, some potential for new smelters to use coal-
fired generation particuarly in the United States
and Australia where large indigenous reserves of
relatively inexpensive coal exist. In other parts
of the world, however, hydroelectricity is by far
the most important power source and accounts for
over 50% of all energy used for aluminium smelting.

123

It is anticipated that a significant portion of increased demand will be supplied either by better utilisation of existing hydroelectric resources or through new hydroelectric developments.

In the glass, brick and ceramic industries, oil and gas currently supply about 85% of energy use and opportunities for greater coal use are expected to be good. However, most of the energy purchased by these industries is used in directly-fired kilns and furnaces where the hot combustion gases are in direct contact with the product. For most glass, ceramic and refractory products, direct firing with coal is not feasible since the coal ash impurities will adversely affect product quality, and the furnaces are not designed for ash-containing fuels. It is likely to take some time before technical improvements in coal combustion can satisfy the quality requirements of some processes in these industries. To date, the emphasis has been on improving the efficiency of energy use in existing furnaces and kilns rather than switching to coal.

In the textile industry where energy requirements are most significant in man-made fibre production and textile finishing, almost three quarters of all energy is produced through steam generation, currently dominated by oil boilers. It is technically feasible and economically attractive for all of this to be supplied by coal-fired boilers. However, the industry is facing severe competition and overcapacity, and most textile and clothing companies have suffered several years of low profitability. In this situation, management is not likely to respond quickly to schemes providing long-term benefits, such as switching to coal, without additional financial incentives.

In the cane sugar refining industry, refineries typically use cogeneration to produce most of their own electricity and process steam for water heating and evaporation. Almost all energy consumed is burned to produce steam in boilers, in which coal could replace oil and gas as a fuel source. How-ever, the industry does not now appear prepared to undertake substantial capital investments for conversion to coal and is likely to stick to the existing boiler stock, as it has experienced financial difficulties in recent years because of increased competition from corn sweeteners, a decline in overall sugar demand, and fluctuating world sugar prices.

In other less energy-intensive industries, the conversion of existing boilers to coal through early retirement is considered less urgent from the economic point of view and, consequently, conversion is likely to be even slower than in energy-intensive industries.

In most OECD countries, the economics of coal-fired boilers are favourable for new factories where the choice is between investment in coal, oil or gas. However, with only modest growth in the demand for steam, growth in coal use in new industries is likely to be slow. Although coal can also be competitive in many cases where boilers are due for replacement, many oil and gas boilers have only recently been installed and there is concern that without some additional assistance, normal rates of turnover would result in only slow displacement of oil in the industrial sector.

5.3.2 Industrialists' Criteria for Fuel Switching

In heavily energy-intenstive industries such as cement, where energy costs could reach half of total production costs, the difference in fuel cost directly influences the competitiveness of a company. Consequently, the reaction of management to fuel cost differences has been quick in these industries resulting in substantial switching from oil to coal. In many industries, however, energy costs account for less than 10% of total costs and do not therefore receive enough management attention or priority.

It has been often pointed out that industrialists generally apply very harsh criteria to what is considered as "non-productive invest-ment". In many cases the required payback period in an investment decision to replace an oil-fired boiler by a new coal-fired boiler is as short as two years. Similar payback periods are often also demanded for other energy conserving investments such as waste heat recovery systems or preheaters, reflecting industrialists' tendency to think that for non-productive investment the recovery of capital expenditure should be shown on the current and next year's budget. Considering the long life expectancy of boilers, a payback period of two years is unreasonably short, and this general attitude toward investment constrains opportunities to expand coal use in many industries.

Another obstacle arises because, after two decades of oil dominance, handling coal is a new experience for many young workers, and the older generation still lives with the memory of coal as a dirty and old-fashioned fuel. Without stronger education and marketing efforts, it is likely to take some time before this psychology changes and ceases to have an adverse impact on switching to coal.

5.3.3 Inadequate Distribution System

Large scale use of coal in the industrial sector will require the development of port facilities, inland transport, storage facilities, and a handling and distribution network. These will develop as demand for coal grows, but the leadtimes required could constrain industrial coal use in the short and medium term.

In Japan, coal handling facilities at ports and maritime transportation have been comprehensively planned. The largest users, new coal power plants, are planned to be built on the sea coasts with their own ports. For the much smaller industrial coal use, the government provides financial assistance to improve coal transportation through the development of coal centres to handle large coal shipments. A 1.5 million ton per year coal centre has been completed and is currently in operation. Other construction plans are to be implemented stage by stage so as to meet the increasing demand. Major improvements have also been made or are underway in Germany, France, the Netherlands, Denmark and Finland. However, some countries have not yet advanced that much in the planning and development stages for new coal handling infrastructure. In Italy, superport locations for unloading have only recently been decided and firm development plans are only now getting under way. Portugal and Turkey need to develop supporting infrastructure but, due to financial problems, no decision on how to secure required investments funds has yet been taken. Lack of infrastructure could become a critical bottleneck for coal use expansion in some countries.

As far as coal storage is concerned, there are port as well as local site-specific problems. Because coal has a lower heat content per unit volume than oil and is in solid form, more storage space is required and different handling systems need to be installed. It is generally estimated that one-third more space is needed for coal than for oil (based on calorific value) since coal requires more storage space, as well as space for ash deposits and preparatory grinding processes. Storage requirements will be particularly difficult to find in congested industrial areas.

To adjust to increased demand by the industrial sector, distribution systems must also be upgraded to handle deliveries to many industrial users. Smaller coal users in particular may require the establishment of a new delivery system, where coal suppliers would assume the responsibility for collecting ash and utilizing or dumping it in a central place. However, the willingness of existing major suppliers to accept full responsibility to assist small users is still in doubt.

5.3.4 Environmental Problems
Coal use leads to noise and dust during mining and distribution, discharges into the atmosphere during combustion, and causes problems of ash disposal after burning. Besides these problems directly related to the coal fuel cycle, longer range and global issues such as acid precipitation and accumulation of carbon dioxide in the atmosphere are receiving more attention. The proper management of pollutants in relation to its costs becomes an important factor for coal use expansion.

6. SYNTHETIC FUELS FROM COAL
Synthetic fuels from coal are expected to be developed only to a moderate extent in the 1980s but could have some impact on coal demand in the late 1990s if oil prices increase significantly, or if strong government support is forthcoming. The initial pattern of development for large unit size coal conversion technologies is likely to proceed on a project-by-project basis.

7. RESIDENTIAL USE
Coal in the residential sector is likely to remain at about the current level. The direct use of solid fuels in the residential/commercial sector is anticipated to decrease toward the end of the century, but the development of district heating could add some coal use in this sector. Increasing solid fuel use in the residential sector is most likely to occur through electricity generation.

8. COAL TRADE
Trading patterns have undergone significant changes in the past two decades, (see Table 17). In 1960, most coal trade was intra-regional and over half of world trade remained within Europe and the USSR. Germany was the major exporter to OECD Europe, and Poland and the USSR were major suppliers to Eastern Europe. There was also a considerable coal flow

TABLE 17

WORLD COAL TRADE IN 1981, HARD COAL
(Million Metric Tons)

To: From:	North America	OECD Europe	Japan	OECD Total	East Europe	All Others[1]	Total Export
Canada	0.1	1.8	10.5	12.4	-	3.3	15.7
United States	14.8	44.8	23.8	83.4	2.4	16.3	102.1
Australia	-	8.1	34.6	42.7	0.4	7.9	51.0
Germany	-	11.5	-	11.5	0.1	0.4	12.0
United Kingdom	-	9.5	-	9.5	-	-0.4	9.1
South Africa	0.7	20.3	4.3	25.3	-	3.9	29.2
Poland	-	10.2	0.1	10.3	7.1	-2.4	15.0
Czechoslovakia	-	1.1	-	1.1	1.9	-	3.0
USSR	-	1.9	1.4	3.3	18.7	-	22.0
China	-	0.2	2.8	3.0	-	3.5	6.5
All Others	-	2.5	0.5	3.0	1.9	1.1	6.0
Total Imports	15.7	111.9	78.0	205.5	32.5	33.6	271.6

Source: IEA/OECD Coal Statistics/UN Energy Statistics/Secretariat Sources

1 Includes statistical difference

from the US to Canada. US exports to Europe (15 million tons) and Japan (5 million tons) accounted for almost all of the seaborn coal trade. Most of the coal trade was metallurgical coal.

Current trading patterns are largely inter-regional and consist of both steam coal (43%) and metallurgical coal. In addition to the US which still controls about 40% of world coal trade, Australia (20%) and South Africa (11%) have emerged as major exporters. Total trade doubled from about 130 million tons in 1960 to some 270 million tons in 1981. Japan has emerged as the biggest coking coal importer (more than 50%) and Western Europe dominates steam coal imports.

The reasons for the sharp increase in trade are largely economical. The US, South Africa and Australia are all low cost producers compared with most of Europe and Japan. When fuel oil prices grew quickly after 1973 and 1979, coal increased its price advantage over fuel oil, leading to coal conversion in Japan and Europe.

Future prospects for the coal trade in the 1980s are mixed, (see Table 18). Following an almost 50% increase in OECD coal trade between 1978 and 1981, imports and exports declined in 1982 and will possibly do so in 1983. Most of the fall in trade is due to the malaise in the steel industry.

Not much growth is projected for metallurgical coal trade during the remainder of the 1980s because of expected low growth in iron and steel output. Steam coal trade, however, is projected to grow rapidly during the economic recovery. The US may remain the largest single coal exporter, but could lose some ground to lower cost producers, in particular South Africa and Australia. The comparative advantage of South Africa and Australia is largely due to the high cost of inland transportation in the United States.

In summary, while demand for oil in the OECD countries is not expected to grow much during the remainder of the 1980s, prospects for growth in coal use are bright. The prolonged recession of the early 1900s coupled with some structural change in energy intensive industries and projected lower fuel oil prices, will prevent coal use from reaching its full potential as projected in the 1980 WOCOL and the 1978 IEA Steam Coal Studies. However, an overall increase in OECD coal use of 30% in the 1980s and another 60% in the 1990s is well within reach. If this will be achieved during the next 20 years, the share of coal in OECD primary energy use

OECD Coal Demand in the 1980s

have grown from about 22% today to 30% by the end of
the century.

Table 18
OECD Projected Coal Trade

(Mtoe)*

	Imports		Exports		Net Imports	
	1981	1990	1981	1990	1981	1990
Coking Coal						
OECD TOTAL	83.36	95.71	80.39	91.66	2.97	4.05
NORTH AMERICA	3.40	4.70	46.61	53.60	-43.21	-48.90
Canada	3.40	4.70	8.76	17.60	-5.36	-12.90
United States	-	-	37.84	36.00	-37.84	-36.00
PACIFIC	47.76	59.00	26.59	30.36	21.17	28.64
Australia	-	-	26.43	30.00	-26.43	-30.00
New Zealand	-	-	0.13	0.36	-0.13	-0.36
Japan	47.76	59.00	0.03	-	47.73	59.00
OECD EUROPE	32.20	32.01	7.20	7.70	25.00	24.31
Belgium	2.95	2.10	0.22	-	2.73	2.10
Denmark	0.01	-	-	-	0.01	-
France	7.15	6.20	-	-	7.15	6.20
Germany	0.45	1.00	6.30	7.00	-5.85	-6.00
Italy	7.97	8.00	-	-	7.97	8.00
Netherlands	2.24	4.10	0.02	-	2.22	4.10
Spain	2.70	3.10	-	-	2.70	3.10
Turkey	0.40	2.37	-	-	0.40	2.37
United Kingdom	1.50	-	0.60	0.60	0.90	-0.60
Other Europe	6.83	5.14	0.06	0.10	6.77	5.04
Steam Coal						
OECD TOTAL	60.18	140.61	43.29	91.40	16.89	49.21
NORTH AMERICA	6.62	6.70	28.66	53.10	-22.04	-46.40
Canada	6.02	6.10	1.21	11.10	4.81	-5.00
United States	0.60	0.60	27.45	42.00	-26.85	-41.40
PACIFIC	9.23	38.90	6.65	32.90	2.58	6.00
Australia	-	-	6.61	32.90	-6.61	-32.90
New Zealand	-	-	0.04	-	-0.04	-
Japan	9.23	38.90	0.01	-	9.22	38.90
OECD EUROPE	44.32	95.01	7.98	5.40	36.34	89.61
Belgium	4.09	5.50	0.33	0.50	3.76	5.00
Denmark	6.52	7.90	-	-	6.52	7.90
France	12.10	11.60	0.49	-	11.61	11.60
Germany	6.92	11.00	1.89	-	5.03	11.00
Italy	5.72	24.00	-	-	5.72	24.00
Netherlands	3.32	6.70	0.61	0.50	2.71	6.20
Spain	2.23	6.40	0.01	-	2.22	6.40
Turkey	-	4.70	-	-	-	4.70
United Kingdom	0.97	-	4.64	4.40	-3.67	-4.40
Other Europe	2.45	17.21	0.01	-	2.44	17.21

Source: IEA/OECD Energy Balances and IEA Country Submissions (1982)

Note: Projections of coking and steam coal trade are in many cases IEA
Secretariat estimates. For further details, see Table C1 for
individualcountries in Part II.

This table is in million tons of oil equivalent. To compare
these numbers with those expressed in Mtoe (million tons of coal
equivalent), the numbers on this table should be multiplied by 1.4

THE CRISIS OF 1983 - PANEL DISCUSSION

The final session of the conference took the form of
a panel discussion of the main issues covered during
the earlier sessions. There were four panelists -
Ray Dafter, Energy Editor of the Financial Times,
Dr Walid Khadduri, the Executive Editor of the
Middle East Economic Survey, Robert Mabro of
St Antony's College, Oxford and Michael Parker, the
Director of Central Policy at the National Coal
Board. In the event they concentrated on the
significance of the March 1983 OPEC agreement and on
the ramifications of the high level of price
uncertainty which still persist. The following
represents a somewhat edited version of the
transcript of their original contributions.

RAY DAFTER
I want to say a few words about the role of
information in the current situation of
uncertainty. I ought to start by pointing out that
sign again, that clear sign of Energy Crisis at the
entrance. For journalists it is marvellous to be
pointed in a direction as clearly as that. Indeed a
crisis is often said to be a problem projected by a
journalist. It isn't a crisis until we have got to
grips with it. The market is clearly influenced by
perceptions and what I am worried about is that the
information making up those perceptions is often
forced. It is information which we have to give
and I think journalists themselves have quite an
important role in the market. Of course, in the old
days the information between producer and customer
passed readily because it happened to be within the
same company. There didn't seem to be much of a
problem; the company itself was able to ease the
supply and demand problems simply within its
organisation. Those were the days when the oil was

controlled by the big corporations. Now control
over oil has, as we have heard so many times, become
fragmented. There are so many more actors
involved. This was brought home to me during the
latest dialogue at the OPEC discussions in London
when we were getting calls daily from OPEC members
and non OPEC members, senior officials of governments of those countries, wanting to know what was
going on. What was the market "feel?" What was the
British government likely to do? What were the OPEC
governments likely to do? There was clearly a need
for some information and people were fishing around
in all sorts of ponds in order to try and get it.
Unfortunately this information is far from perfect
and I will just mention a few examples. Firstly, we
have no really good data on the stock position of
oil companies, or their de-stocking activities.
That makes it very difficult to put stocks into the
supply and demand picture. Secondly, we are forced
to be more sceptical about the spot market on which
we become more and more reliant. The spot market
has become so very important for the oil trade. And
yet you look at the number of non-existent cargoes
which have been traded in the North Sea - the Brent
crudes for instance. These are in reality no more
than gambling exercises. Those are the prices that
are being reported by ourselves and other people and
they are having some influence on the spot market.
Thirdly, we are unsure about the production and
exporting policy of the Soviet Union. The data
tends to be late and hazy and we have to try and
make up to date prognoses of what is going on.
Until last night I think we were uncertain about the
British government's own position regarding the UK's
role in the world oil market. I agree with Edith
Penrose, that was a very significant statement by
the Energy Secretary saying that Britain has a role
to play on the world stage of oil.
       What then are the things which would help us
get a better handle on what is happening? One is
the need for a greater dialogue between producers
and consumers. This cannot be formal because the
market doesn't work like that. It is not only a
case of OPEC and the OECD because, as we have heard,
there are so many more actors now in the market.
But clearly, the more information we have passing
between the two sides the less uncertainty and the
less chance of violent swings we are likely to get.
Secondly, there needs to be more accurate
information on the spot market transactions. There
are publications which are not so much reporting

what happened yesterday, but actually making up
prices which are influencing the spot market. There
needs to be a better reporting service there. I
look forward to the futures market developing in
both crude oil and, of course, products to give us
greater clarity. Journalists coming to conferences
such as this also help because the better informed
we are the more accurate the reports will be.

WALID KHADDURI
I would also like to talk about uncertainty, but
about the politics of it. If we compare the two
decisions of October 1973 and March 1983 we find
that both market forces and political circumstances
were involved. The October 1973 decision was the
response to continuously rising demand for cheap
oil. It was implemented by the actions of Libya
three years beforehand, and also by the behaviour of
the US in providing assistance to Israel during the
1973 war. In 1983 the situation was different.
Demand was shrinking, the energy cake was smaller
and OPEC had to concentrate its mind and take a
decision to retract its nominal price. Politically,
we find that the OPEC decision in 1983 was cemented
by the co-operation between OPEC as a whole and the
western countries, in particular between the UK and
a few OPEC states. During these past ten years a
change has occurred in the quality of OPEC
decisions. Whereas on several occasions in the past
OPEC took decisions, but did not hold to them, as
for example in the case of the Qatar Agreement, or
of the March 1982 agreement, this time when collapse
was imminent - and I think that if there was no
agreement in London in March oil prices would have
collapsed - they decided to agree, despite all the
big political differences amongst the sovereign
states. Again we find that after ten years of
dialogues, seminars, new world economic order, the
Paris meetings of 1975 - which did not really result
in anything - when the crunch came there was an
agreement between OPEC as a whole and Westminster,
as well as with Mexico.
     Now what does this tell us about the future?
I think the consensus at this conference is really
that nobody knows. The future is uncertain. How
will the OPEC states act in the future and how will
relations between producers and consumers develop in
the next few months or years? In the Middle East I
think we are going through the most fluid and
dangerous political situation ever since World
War II. The few safety valves that remain are very

tight and any one of them can blow at any time. We
have several wars going on and there is almost a
stalemate in taking any decisions. If in these
circumstances anything goes wrong, the impact on the
oil situation and on the oil industry would depend
on the circumstances and the type of conflict. We
also have the economic instability now in several
OPEC states including Venezuela, Nigeria, Iraq and
Algeria. We have it in non OPEC states, as well,
Egypt for example. The Gulf states, with surplus
money, are going .to have their own problems also,
arising from the fact that they have to accommodate
themselves to lesser revenue. I think we will have
to wait and see whether the political decision
making processes can accommodate this development.
It is not easy to accept lesser revenue. Many
people in the area want lesser production and less
revenue so that we can focus our minds on better
rational decision making. It remains to be seen
whether that is possible or not. All of these new
circumstances that we are going to face will have
their impact on our own local oil industry,
investment in the oil industry, oil production
policy and will eventually also effect oil
consumers.

ROBER MABRO
The main structural features of the oil market have
not changed at all as a result of the OPEC agreement, or the BNOC decision on prices. We still have
today a "market"; there is enormous excess capacity
in production, particularly in the OPEC countries,
although not so much elsewhere. The number of
players, it has been said ad nauseaum at this
conference, has increased. On the key exporting
side, the number of exporting countries has grown -
some of them are very important and not part of the
existing club. On the industry side, the number of
barriers has increased and the mode of transaction
has changed - people are no longer buying their oil
on long term contract, but rather at arms length on
short term contracts. In these circumstances the
market only stablised when there was a realisation
among the countries in OPEC that if the price of oil
continues to fall it would be almost impossible to
stop it. This realisation came about following the
actions of the Nigerians and of the Saudis. What
the Nigerians promised was that if BNOC lowered its
price, they would match it cent by cent. They have
thus brought around the discussion table the concept
of the price war in a concrete form. It was no

of the price war in a concrete form. It was no
longer the abstract idea of academics but a
political reality. When the Saudis for the first
time in the history of OPEC said that the market
should come down by $4.00, the producers realised
that if the price were to remain at the lower level,
there would be a substantial loss of revenue in the
short run. That realisation - I call it the fear of
god - concentrated the mind and led to an agreement
which overcame all the political resistance and all
the differences in political ideology.

The question which has to be asked for the
future is how long can this agreement last? I think
that if in 3 or 4 months there is no change in the
real force of demand and supply then competitive
instincts will re-emerge. If, however, demand
begins to rise, it will serve to support the will to
maintain the price. Likewise if supply contracts as
a result of a new Middle Eastern war, for example,
there will be a hardening of producer attitudes.
But if neither of these things happen, then it is
unlikely that the present level of agreement would
continue.

MICHAEL PARKER
Well ladies and gentlemen, I come here as a member
of a discredited band on two counts. First of all
as a member of the bureaucracy of a nationalised
industry, and secondly as some kind of energy
economist. You may have noticed that the energy
economists have moved away from forecasting to
scenarios which, in my view, is an act of corporate
cowardice. It does reflect, however, what has been
happening in a wider scene. In times of shortages,
governments and consumers want planning. In times
of surpluses they want the market to reign supreme.
This is, of course, only human nature. The problem
with this transition has been to know what running
the energy sector by market forces really means.
In the majority of cases it seems to mean that all
decisions are to be taken on the basis of current
prices. Yet it is manifestly the case, as Professor
Hafele has been showing today, and others no doubt,
that it is price expectations and not today's prices
that matter for most of the structural things that
we have to do. Certainly this is true of coal.
Thus I do not believe that very much of the
increased use of coal since the first Arab shock of
1973/74 has been attributable to the investment
decisions taken since 1973/74. Most of the growth
in coal demand will have occurred as a result of the

spare capacity to use coal arising from decisions taken before 1973 - some of them perverse decisions at the time. Uncertainty over price expectations means that it is going to be harder to create demand for coal than to create supply during the next ten years.

Another factor which is affecting coal, as with oil, is the increased importance of the spot market. The spot market is, however, extremely volatile and it is this volatility which has an influence out of all proportion to the spot market's physical dominance. It has led to the erosion of peoples' belief in planning. I fear that the robust strategies which are developed in those circumstances - that is everybody planning for the worst eventuality - will actually ensure that we get the worst eventuality. The existence of market uncertainty on the one hand, and the need to take long-term action on the other, presents a major intellectual challenge which has not yet been met.

Printed and bound by CPI Group (UK) Ltd, Croydon, CR0 4YY

21/10/2024

01777088-0020